SUNBIRTH

ALSO BY AN YU

Braised Pork
Ghost Music

AN YU

Sunbirth

HARVILL

1 3 5 7 9 10 8 6 4 2

Harvill, an imprint of Vintage, is part of the Penguin Random House group of companies whose addresses can be found at global.penguinrandomhouse.com

Vintage, Penguin Random House UK,
One Embassy Gardens, 8 Viaduct Gardens, London SW11 7BW

penguin.co.uk/vintage
global.penguinrandomhouse.com

First published by Harvill in 2025

Copyright © An Yu 2025

An Yu has asserted her right to be identified as the author of this Work in accordance with the Copyright, Designs and Patents Act 1988

Penguin Random House values and supports copyright. Copyright fuels creativity, encourages diverse voices, promotes freedom of expression and supports a vibrant culture. Thank you for purchasing an authorised edition of this book and for respecting intellectual property laws by not reproducing, scanning or distributing any part of it by any means without permission. You are supporting authors and enabling Penguin Random House to continue to publish books for everyone. No part of this book may be used or reproduced in any manner for the purpose of training artificial intelligence technologies or systems. In accordance with Article 4(3) of the DSM Directive 2019/790, Penguin Random House expressly reserves this work from the text and data mining exception.

Typeset in 12/16.62pt MinionPro by Jouve (UK), Milton Keynes
Printed and bound in Great Britain by Clays Ltd, Elcograf S.p.A.

The authorised representative in the EEA is Penguin Random House Ireland, Morrison Chambers, 32 Nassau Street, Dublin D02 YH68

A CIP catalogue record for this book is available from the British Library

HB ISBN 9781787304970
TPB ISBN 9781787304987

Penguin Random House is committed to a sustainable future for our business, our readers and our planet. This book is made from Forest Stewardship Council® certified paper.

All that is pure lies in the primordial and the divine, everything in between is filth.

SOMEWHERE, SOMETIME, A LONG TIME AGO

The Sun People stripped the blind man of his clothes
    and ordered him to walk.
'Find the Sun,' they said. 'Find the God of Light.'
'There,' they told him. 'You will finally see.'

Naked, the blind man marched towards the Sun.
Three hundred mornings, he treaded east.
Three hundred dusks, he hiked back west.
Never once did he reach the Sun.
He walked in darkness, until he no longer believed
    in the light.
The world must've always been black.

He lost his faith,
and was left with a limited body, wrapped around
    his bottomless desires.
Those desires grew.
His body remained.

The Sun People gathered around as the blind man
    fell to the ground.
'Help me,' he begged. 'I am but a man.'

But silently, the Sun People stood and watched,
    as a light, a brilliant gold, poured out of the blind
    man's mouth.

The light grew bigger,
and swallowed the blind man's head.

Upon witnessing what had happened, the Sun
    People picked up the blind man's body and built
    him a temple.
'He found the Sun!' they exclaimed. 'He found
    the Sun!'
'The Light is in him!' they shouted.

As the years went by,
as mountains eroded,
as rivers broke the soil,
the Sun People lost their God, the sun,
and believed, now, in the Blind.

The sun dimmed and disappeared.
Everything died.
The Blind died, too.

No one lived to tell what happened thereafter.

FIVE POEMS LAKE, A LONG TIME THEREAFTER

1

The sun was half bright, half warm, half full.

It was a morning in August and it was cold. I'd made the mistake of using a damp towel to seal the gap under the bedroom window, and now the fabric had frozen solid. I must've caught a cold, because I'd woken up with a stuffed nose and a painful lump in my throat. After having rubbed my legs as fast as I could for a few minutes, I got out of bed and boiled some hot water on the stove. I unclasped my lab coat from the rack, relieved to discover that it was finally dry. The lab coat had felt too tight for some time now, ever since I started wearing my coat underneath, but I didn't care enough to replace it with a larger size. Dong Ji had reminded me on many occasions that I looked like a gigantic steamed bun. She couldn't understand why I never felt embarrassed looking so silly.

I'd hung up some quilted curtains on the door connected to the shop floor, but they hardly did anything to keep the draught from creeping into the living area.

The colder and darker mornings nowadays meant that I'd been having trouble waking up at the usual hour, leaving me with just enough time to quickly stuff down a boiled egg before opening the pharmacy.

To soothe my throat, I crushed up some dried monk fruit and dropped it inside my hot water flask. I went into the shop, unbolted the lock on the front door and turned the wooden sign to the 'Open' side. There were already two people waiting by the steps: a balding man and a short, plain-looking girl whom I recognised as one of the Su siblings. The man was dressed in heavy jeans and a brown waxed coat. All the angles on his face – his nose, chin, brows – were so pointed that I imagined his teeth to be sharp as well. I bade them good morning, to which he responded with a puzzling, 'We're not together', and proceeded to hand me a prescription that he'd been holding.

I took the paper from him and looked towards the east. The morning sun was like half of an orange, rays oozing out from its cut side.

'We've made it twelve years,' I said.

'I don't want to talk about it,' the balding man mumbled as he gestured for me to lead him inside.

The first time a part of the sun disappeared was on the fifth day of August, exactly twelve years ago. It almost went by unnoticed. If it wasn't for a schoolteacher who was measuring the sun with her students, it really could've passed like any other rotten day. It makes you wonder whether the astronomical observatory is just a place for researchers to sleep, though eventually they too discovered that the sun had risen like usual that summer morning, but a sliver of it just wasn't there. It was like a waning gibbous moon, they said at first, but it wasn't long before they stopped calling it that, because, unlike the moon, that sliver of the sun never came back. Since then, from time to time, entirely unpredictably, a little ribbon of the sun would vanish again. As this happened, days became colder, winters longer, birds flew away, leaves turned black and the lake froze over.

There wasn't one person at Five Poems Lake who could tell you where the other half of the sun had disappeared to. Maybe it'd abandoned us for another town – if such places still existed – but we had no way of knowing that. There was a time when people cared, gathering in front of the astronomical observatory, demanding answers. Some who believed too much in their capabilities took it upon themselves to come up with solutions, but it soon became clear that the sun was not something we could understand or control. By the time winter fell upon us, having discovered no new information, the head researcher of the observatory resigned, putting an end to the protests. Despair blanketed the town, days went on, years slipped by and nobody asked questions any more.

Now, you might ask, how could they stop there? Half of the sun was gone.

Beats me. If I had to guess, I'd say that it was because of time passing. Anyone who spent twelve years under a fading sun had better accept it as part of the natural course of things. People had to live, after all, and life was already a large enough mess, even before the sun started to disappear. This may all sound foolish – mad, even – but as long as the sun continued to rise, we had to live, and to live, for most people, was to survive as an individual, not as a species. So, time. That is my guess. Time stretches everything thin. Change is a creature that crawls.

I guided the balding man to the back counter, reading the prescription while I walked. He followed keenly, so close that I could feel his warm breath on the crown of my head.

'We're out of wu wei zi,' I said, relieved that the counter was separating us now. 'It's been weeks since we had any stock.'

The man's hands were rubbing against each other, but they went limp and slid onto the walnut countertop.

'When will you be getting some more?' he asked. His nasal voice paired well with his facial features.

'It's hard to say. Maybe next month if we're lucky. All the suppliers tell me the shortage is only going to get worse. But you never know!'

I tried my best to sound optimistic.

'Do you have everything else that's listed in this prescription?' he asked.

'I do, but it might cost you a bit more than usual.'

He began deliberating, so I folded the prescription and gave it back to him. The Su girl was standing in the corner, torso leaning over the counter, neck stretched forward, squinting at the small labels on the apothecary cabinet. She was wearing a white denim jacket – too little for weather like this. There was a white scarf tied around her neck. Her coarse black hair was knotted into a single braid that ended where her legs began.

'What can I get for you?' I walked over to her.

I'd never spoken to her before. She had a limp in her right leg that had been there for as long as I can remember. All I knew about her was that her family once owned the boat that used to shuttle people across to the other side of the lake before it froze.

'I was just looking,' she said. 'You need new labels. The writing is all faded.'

She spoke in a strange manner. It wasn't her tone but rather the way she pronounced her words. It sounded too crisp, I suppose.

'I've been putting it off,' I said. 'My great-grandfather made those. It feels wrong to replace them. Plus, I know where every ingredient is.'

Her skin was pale and her lips were so thin that they seemed as though they were constantly pursed. Up close, I realised that she was younger than I'd thought. Sixteen, maybe seventeen. Her expression had a maturity that masked her real age. She pulled her scarf up to her nose and took a sniff.

'Some of this stuff can be poisonous if combined the wrong way,' I warned her. 'I suggest seeing a doctor before you buy anything.'

One of the regulars, Miss Pan, walked through the entrance and waved her gloved hands at me. She was wearing a mauve coat that had circular holes cut out from the fabric. Underneath, she had on a second, bright red coat. She looked like a flower.

'You're welcome to stay and look around,' I told the Su girl. 'I'll be here if you need me.'

'Good morning, Miss Pan,' I said. 'You're early today.'

'Look,' Miss Pan said. 'No one should sleep in on a Monday. Not even me. It's important to start the week off right.'

Her voice rang across the shop. Every time she came in, she would fill the place with her presence. Gracefully, she removed her leather gloves from her hands, loosening each of the fingertips before pulling the whole thing off. Her pregnant belly was bulging behind her two coats, and she pressed it against the counter like a cushion.

I squatted down, my blocked nose making my head spin, and unlocked one of the lower cabinets. I pulled out the two boxes of bird's nests she'd ordered.

'I bet you've never heard this before,' Miss Pan said, shooting a quick glance at the Su girl, 'but my cook is terrible at cooking.'

I laughed and wiped the dust off the boxes with a tea towel.

'I don't know why I keep eating her food,' she continued. 'Maybe I do it because she's worked for my family ever since I was little, but for heaven's sake, her rice is like congee.'

'There are cooking classes she can take,' I said. 'Did you see the sun on your way here?'

'Of course, silly girl, it's in the sky.'

'That's not what I mean.'

'Then I don't know what you mean.'

'It's been twelve years since it started disappearing,' I said. 'Did you think we'd make it this long?'

'Back then, I felt that just making it to the next day was a miracle.'

'Do you feel differently now?' I asked.

She looked down at the bird's nests and answered my question with a soft hum.

'Looking back at it all,' I said, 'we seem to have done all right.'

'You think so?'

She grinned at me and flipped one of the boxes around to read the back label. She had silken hands, much like my sister's, milky and fine.

'I have an idea,' she said. 'Do you think you can cook this for me? I'll pay you.'

'I don't run a restaurant, Princess Pan. It's not even that difficult to make.'

On occasion, I liked to call her that. Princess Pan. She must've found humour in it too, because she would lift her chin up slightly and show her little nostrils to me. Sometimes, she'd hold out her hand and I'd pretend to kiss it or spit on it, and no matter what I did, both of us would laugh.

'I'd take it to a restaurant, but they always use too much sugar,' she said. 'Help me, will you? All you have to do is throw it in a pot with some water.'

'You've never cooked a meal in your life, have you?'

She gave an embarrassed smile as she shifted her weight onto her other leg and pushed the box back to me.

The balding customer seemed to have made up his mind. He walked up behind Miss Pan and tried to get my attention by waving his prescription in the air. I looked around. The Su girl had left at some point without my noticing.

'All right,' I told Miss Pan. 'Just this once. I'll get it delivered tomorrow to your home.'

'You've just made my day. Cook one box and keep the other one for yourself. Think of it as a thank-you gift from me. Here, take these too. They're quite soft.'

With her slender fingers, she combed through the fur on her gloves and then handed them to me.

'I'll ban you from the shop if you're going to keep doing this,' I said, pushing her hands away. 'I haven't even worn the hat you gave me last time. Keep these.

It's cold out. I'll store the other box of bird's nests for you until you're done with the first.'

She reached effortlessly over the counter and stuffed the gloves into my lab coat pocket.

'If you don't want them,' she said, 'give them to Dong Ji. Now, how much do I owe you?'

I wrote her an invoice. She handed me some bills and insisted that I kept the change. She was always forcing me into taking things – expensive clothes, money, food – which made me feel closer to her and farther from her all at once.

'Thank you,' she said before she left. 'Really.'

I waved goodbye to Miss Pan as she walked out the door, put the bird's nests back into the cabinet and turned my attention to the balding man.

'I'll take it,' he declared. 'Everything you have.'

'Everything I have?'

'Your entire stock of the ingredients in this prescription,' the man said, stabbing at the piece of paper with his index finger. 'I'll take it all.'

It wasn't uncommon for customers to offer to buy more than they required. People were afraid that things would run out, so they'd snatch up all sorts of medicine they didn't need and had no idea how to use. I always declined their requests, of course, because I couldn't run the shop without a balanced inventory. I wouldn't be able to fill any prescriptions that way. On top of that, it was a waste of money most of them didn't have. Yet I must admit that the fading sun had been, in a way, a blessing for pharmacies.

'I can't do that, sir,' I told him.

'Why not?' he said, shocked. 'It's cash for you. What's the difference? This medicine is really important to me.'

'Sir, just come back once you run out and I will refill your prescription. This summer's been colder than the last. There are plants that just don't grow any more. I need to make sure all my customers can find what it is they need.'

'But you don't even have all the items I need now!'

'I'm expecting a delivery tomorrow morning. Maybe I'll get some wu wei zi. You can come check again then.'

'So sell me your stock now. You'll get more tomorrow, right?'

'I can't do that, sir. No matter what you say.'

'Well, you've just lost my business.'

He snatched up his prescription and stormed out the door, stirring up a puff of dust. The balding spot on the back of his head was shaped like a keyhole.

Of everything that had vanished over the past twelve years, I did not miss much. Five Poems Lake had been in decay long before the sun began to disappear. Buildings had been deteriorating, machines breaking down, soil eroding, the population ageing. I did find myself thinking fondly of the old seasons, despite the fact that we'd never had much change between them. Five Poems Lake had been a hot, humid place, though it was surrounded by the endless desert. Winters were just cool enough to indicate a shift towards a new year: when those days arrived, we wore skirts with bare legs, winter jackets on top, wool hats. We sucked on popsicles and drank hot ginger tea. These contradictions made sense to us. Girls would wake up early every day to put on make-up, showing the world their long lashes, cherry lips and velvety cheeks.

We spent more time outside than at home.

In contrast, summer days were simple. Life was defined by sweating. Clothes stuck to skin, hair wet, bare feet slipped in sandals. We stored away our eyeshadows and foundation, opting for naked and honest faces, as though with the summer sun watching from above, we had nothing to hide. We knew the heat well. The sun felt close to us.

As people grow up, childhood becomes foggy, but no matter who you ask, they always remember the weather. They forget the locations of their schools, the names of their best friends, the colours of their houses, but never the weather. Perhaps it is the tactility of air touching skin that makes it impossible to forget. We remember it with our bones, our muscles, our skin. So even though I had been living with the cold for over half of my life, if I just closed my eyes, I could still feel the full sun blazing over Five Poems Lake.

The Su girl came back later that day, just as I was about to close the shop. The soreness in my throat had eased a little from all the monk fruit tea I'd been gulping down, but my nose wasn't any clearer.

'Did you go see a doctor?' I asked her.

She shook her head.

'Are you looking for something for your leg? Is the cold weather bothering it?'

Again she shook her head. 'My knee got smashed by a rock when I was six. I hardly feel it any more. It's like it's not mine.'

She looked like she was going to say something else, but the words grazed those small lips of hers for a moment and slid right back down her throat.

'What can I get for you then?' I asked.

Through the door and across the lake, the red sun was slipping among the buildings in the west, like it was searching for a secret place to sleep. I imagined its other half, split into pieces, each hiding somewhere in the west. Under tables, in garbage bins, in toilets.

What I find ironic is the fact that the lake is shaped like a temple bell. It is rumoured that a long time ago, when those who constructed the tall buildings and engineered the cars were still around, when we were connected to some greater world, an esteemed monk named Shan Qi travelled here on his pilgrimage and fell in love with a woman. He decided that he would write four poems for her, one for each season. If she accepted his love within the year, he'd resume secular life and marry her. He spent months composing these poems, but even after he'd presented her with all four, she refused to be with him. In his despair, he surrendered to his desires and wrote a fifth poem. It was another about winter, the real one in his home towards the far north. There, it was cold and bleak and quiet – the kind of winter the woman had always dreamed of but never experienced. She was moved by it. She accepted Shan Qi's love, he renounced his life as a monk and together they settled down.

Sometimes, when the sun was hidden behind clouds, when the nights were a few degrees too cold, when my joints hurt from the wind, I would find myself wondering whether it was that woman's fault that this was happening to us. I am not one to be superstitious: I pride myself in believing empirical facts, things that prove to be correct time and time again. Such an attitude is necessary for those in my profession. Yet, though I tried to fight it, I couldn't help but think that maybe it was the woman's dream, her yearning for another kind of winter, that drove away the warmth we'd once enjoyed. Maybe it was the monk who brought the northern winds with him. Maybe it was the poem he wrote. That fifth poem, born from one man's failure to control his desire, next to this lake that was supposed to ring with faith.

Though I have to say that the tale seems rather far-fetched. There were no travellers here. Everybody who lived in Five Poems Lake was born here and had stayed.

There was no one to stop us from leaving, only the endless desert. Nestled among arid and desolate sands, the lake was regarded as a miracle. That is not to say that people didn't leave. Many did, but none of them returned. Perhaps they never made it to another place. Or they did and saw no reason to come back.

The setting sun crouched on the horizon like it was wounded. Its last rays fell over the steps in front of my shop, painting different shades of amber onto the uneven bricks. The Su girl pulled her white scarf up to her face.

In the way her facial muscles stretched and relaxed, she still very much resembled the child that I'd sometimes seen sitting in the corner seat on her family's ferry, drawing in a notebook. But her bodily movements were no longer the same; they'd become heavier. Perhaps it was because of the leg injury. Or perhaps it was because she was no longer a child.

'I love this perfume my brother gave me,' she said.

She untied the scarf and waved it in the air in front of my face. It smelt like fresh pear, or maybe melon. Both were scents I hadn't come across in a long time.

'It's refreshing,' I said.

She smiled, gave it another sniff and wrapped it around her neck.

'Does seeing the sun make you anxious?' she asked, noticing that I was looking outside. 'You seem like you don't know whether it'll rise again tomorrow.'

'We don't know, do we?' I said. 'But we can't control what happens.'

'See? I don't know how you can think like that. When I feel anxious, which happens all the time, by the way, I get myself to think about aliens. It's the only thing that makes me feel better.'

'Aliens?'

She crossed her arms over the counter and leaned forward.

'I like to imagine that there are others out there,' she said. 'Maybe they are aliens who live with a different sun. Maybe they've adapted to surviving in darkness, like ants or something. When I think that way, it makes me feel like anything can happen. It's possible that the sun won't rise tomorrow, but it's also possible that we will be the ones to disappear before the sun does. What if we just vanished like smoke in the middle of the night? My leg might have been smashed by a rock, but once I could've walked across the desert to another place. When I open up the possibilities like that, I feel like nothing's personal any more. I'm like a leaf drifting through the sky.'

'You could get lost in the stars, or you could end up on the moon? You might even find the parts of the sun that disappeared.'

She nodded. 'You're quick on the uptake.'

'Thanks for the compliment.'

It was getting dark inside. I turned on the light

switch. The ceiling fan light flickered for a moment before it stabilised. I'd been wanting to change the fixture. The light attached to the fan wasn't bright enough, and the fan hadn't been used in years.

'Does Miss Pan come here often?' the Su girl asked.

I nodded.

'Was she wearing two coats?' she asked. 'Seems silly.'

'You can hardly blame her for that, can you? It's freezing.'

'I saw her car parked outside.'

'The black one?'

'Yeah. Have you been in one?'

'What?'

'A car.'

'Oh, yeah. I've been in police cars. My dad was a police officer. Plus, buses are pretty much the same thing, just bigger.'

'That's my dream.'

'What? To be a police officer?'

'No way!' She laughed. 'To drive a car. *My* car. A car I own. I love cars. But I don't know if I'd be able to drive. You know, with my leg.'

'You can use the other one.'

She looked down at her leg, giving my suggestion some serious thought.

'I wonder why no one has tried to drive a car out of here before,' I said.

She raised her head all of a sudden and looked into my eyes. I recoiled a little.

'Because there are no roads out,' she said. 'You can't drive on sand.'

She began massaging her bad leg with one hand.

'When is Miss Pan's baby due?' the girl asked.

'Soon, I think. It's hard to believe.'

'What is?' she asked.

'That there are still people with the courage to bring a child into this world.'

I looked at the clock on the wall. It was time to close the shop. I began arranging the invoices from the day into a neat pile.

'But Miss Pan has always been that way,' I said.

'Been what way?'

'Oh, you know. Brave, I suppose. Able to see the joy and hope in things.'

The Su girl tied her scarf into a bow and then pulled the knot open again.

'So what did she buy?' she asked.

'You mean when she was here?'

'There were some boxes. I saw you grab them for her.' She reached her hand over the counter and pointed at the lower cabinet.

'Oh, she bought some bird's nests.'

'Must be expensive,' she said.

I nodded.

'You think she bought them precisely *because* they're expensive?' she asked.

'Maybe, but why she buys what she buys is really none of my business,' I said. 'As long as I can sell my stuff, I'm happy. I stick to my end of the transaction. Anyway, if you want any, I'd have to put in an order for you. But, as you already know, they cost quite a bit.'

'Someday, maybe. If I ever get the money, I'd love to try some.'

Though she had no reason to stay, she seemed hesitant to leave.

'What do you need the bird's nests for?' I asked, to fill the silence. 'To be entirely honest with you, I don't know if they're really worth it.'

She grinned at my question. It surprised me to see that her mouth stretched so wide.

'It's kind of like driving a car,' she said. 'I've just always wanted to try it.'

'Do you know what kind of birds these nests are made by?'

She shook her head.

'Swiftlets. Or Golden Silk Birds. We never had many of them here.'

She chuckled.

'What's so funny?' I asked.

'Golden Silk Birds,' she said, still smiling. 'Of course they make expensive nests. It's a shame that they don't like to live here.'

'They like to live in caves, and we don't have many of those. All the pharmacies are fighting over the last of the bird's nests.'

The Su girl seemed disappointed. She sighed through closed lips, making them look fuller for a second.

'I get loose bits of nests sometimes,' I said, trying to cheer her up. 'I'll remember to keep some for you. They're cheaper.'

She thanked me, and with her hands stuffed in her pockets, she walked out once again without buying anything, her bad leg dragging behind.

I opened a box of bird's nests and emptied it into a bowl of water to soak overnight. Then, I took a quick bath in the wooden tub that our grandfather had built when we were young. The steam cleared my nostrils for a moment, but the water cooled too quickly for me to stay in it for long. While the bath drained, I burned some incense for Ba. I used to be more diligent, but recently I would often find myself in bed, half caught in a dream, before realising that I'd forgotten all about the incense. It'd be too cold to get out from under the duvet, so I'd leave the task for the day after.

The orange on the altar had grown some green mould, so I replaced it with a green apple. If I were honest, I had been deliberating over taking away the fruit and replacing it with a bowl of rice grains. Why would the dead need to eat fruit? Especially when anything fresh had become so scarce. I cut the good half of the mouldy orange into three slices and sucked on them as I brushed the wood with a feather duster. When I was done, I straddled a plastic stool and began drying my hair with a towel.

'I thought I'd spend some time with you tonight,' I said to Ba. 'Just until my hair dries. I have to get up early tomorrow. We're getting a delivery. I never liked waking up early, but now it's even harder.'

Our father died when I was ten. My sister Dong Ji was in high school. It happened just a few months after the sun disappeared for the first time. When I came home from school, our grandfather had a white porcelain urn in one hand and our father's police badge in the other.

'Your father is gone,' Yeye told me.

Half an hour later, my sister walked through the door and Yeye said the same thing to her.

Yeye took it upon himself to inscribe our father's name onto a wooden placard. After spending two days in his room, eating nothing and speaking to no one, only showing his face to use the toilet and fill his water bottle, he finally came out with a sandalwood placard about the size of a man's shoe. There was no 'rest in peace' or 'in memory' inscribed on it. Nothing of that sort. Just 'Dong Yiyao' in big golden characters. Even though it was spare, I really thought our grandfather did a fine job. Every stroke was smooth like it had been painted on with a brush. He must've carved one of these before, I thought, maybe for the grandmother I never met.

During the two days and nights Yeye worked on that placard, Dong Ji and I sat around the porcelain urn and discussed whether we should take a look inside. We were too afraid, we decided in the end, though I couldn't tell you what it was that we were afraid of.

After the placard was finished, the three of us took a bus to a woodworker near the north of the lake and ordered a simple altar for our father. The bus swayed steadily as it drove past stretches of farmland and into an area filled with warehouses and workshops.

Our town was built all around the rim of the lake. The western edge of the water, along the crown of the bell, was where all the tall buildings were, and outwards from there, the buildings become lower, older and sleepier, as though the lake itself was a time machine. The east of the lake – the mouth of the bell, if you like – was a quiet area of bunched grey-brick buildings, most of which were just one storey high. The police station was the only exception. It towered over the area. We always joked that it was the palace of the Eastern District, where emperors and princes spent their days snoozing with their concubines. My father, an earnest police officer who worked hard all the time, never liked the joke.

We lived in an old family home in the Eastern District, right next to the police station. Our house was separated into two areas and therefore larger than most of the other homes nearby. The front was a herbal pharmacy that my great-grandfather built, and the back was where I now lived alone.

The altar we picked was the colour of chestnut and just big enough to fit the placard, two candles, an incense burner, a baijiu glass and a single piece of fruit. Or, in the concise words of the carpenter, 'all the essentials'.

Unlike Yeye, who had already purchased himself a burial plot next to our grandmother, Ba had always insisted that he didn't want to end up in a cemetery.

'When I'm dead,' he'd often said, 'I don't want a funeral. Nor do I want to be buried. A dead man should be allowed just to be dead.'

As much as he would've liked it, we couldn't just dump him into the garbage. We had to come up with something else. Dong Ji suggested scattering his ashes into the lake but, thinking of the children who spent their summer days swimming in the water, we all agreed that it felt wrong. Yeye came up with the idea of setting the urn in a pot and planting something over it. Ba had always loved plants, so this seemed fitting.

Again the three of us went out, this time to the flower market. We picked out a camellia and a deep terracotta pot. Yeye pushed the pot and the plant in a trolley while Dong Ji and I each carried bags of soil in our backpacks. We put the urn in a plastic bag, placed it at the bottom of the pot, poured in soil and transferred the plant over. We dragged the plant to the entrance of the shop so that it would be the first thing people noticed when they walked in. We decided that our father would've enjoyed all the company. The camellia ended up standing rather tall, lush green, with pink flower buds sprouting all over. I remember deciding that I wanted to be buried in this way too.

Nobody knew why our father was at the lake the night he died. He often spent days at the police station without coming home, so Dong Ji and I did not think too much of his absence. We were told later that they found his body washed up on the shore with no signs of a struggle but Yeye wouldn't let anyone see it. Yeye took care of it all by himself – the paperwork, the placard, the funeral. He didn't even tell us until the funeral was over. During the weeks after Ba's death, there would be extended moments in the day when nobody talked. We didn't know what to say. Yeye had taken me to someone else's funeral before, which had been the only time I'd seen people grieve. The first half had been so quiet, it made me think that grief was silent. Later on, everybody started crying, so I thought that maybe grief was a cacophony of sobs and broken sentences. It wasn't until those days after Ba's death that I came to learn that it is really none of those. Grief is a running tap, a closed cabinet, chopsticks touching bowls, teeth chewing. I'd never imagined that would be the case. I hadn't expected it to slip into everything.

Even Ba's team didn't know what really happened. At the time, Ba had been working on some missing persons cases. Three people had disappeared within the span of a week. Two were married, and the other was a university student who had nothing in common with the couple. The details of the cases were kept confidential.

His team must've felt responsible, perhaps because of something related to the case that we weren't told, or some special kind of bond between police officers that I couldn't understand. They came by our shop one by one and gifted us all kinds of things – flowers, fruits, alcohol, biscuits – saying that they wished they could've prevented it. Some even kneeled in front of the camellia and cried. Our grandfather was too proud – too kind – to demand answers, so in those two days he'd spent alone in his room, etching the big characters onto a piece of wood, stroke after stroke, he'd quietly accepted his son's death.

A headache woke me up at four-thirty in the morning. Going to sleep with wet hair always gave me a dull pain in my temples. Dong Ji had the same problem, which she'd warned me about years ago, but I'd never learned.

I took some medicine for the pain, sat on the toilet

and squeezed the towel that was hung on the door. It was coarse, thin and still sopping from the night before. Useless old thing.

As I waited for the sun to rise, I steamed the bird's nests with some almond milk, goji berries and rock sugar. At seven o'clock, I turned off the stove and stood in front of the television while brushing my teeth. The local weatherman was talking about the sun. He still always matched his tie to the season. Spring was green, summer was orange, autumn was red, and winter was blue. Today, a dark orange tie with white stripes was knotted neatly around his neck. I wondered whether the colour choice was made out of optimism or denial.

'It's looking to be a very cold summer day here at Five Poems Lake as the sun has shrunk some more this morning,' said the man in a severe tone. 'After almost two months of inactivity, it looks like this fading is the most substantial we've seen so far, measuring over eight per cent. At this point, the only thing predictable about the disappearance of the sun is that it will happen.'

Outside the window, the streetlights were still on and the sky was mostly dark. Hurriedly, I finished brushing my teeth, wrapped myself up in some extra layers, and joined the people who were already standing outside, heads stretched towards the east, as though they wouldn't believe it unless they confirmed it with their own eyes. The clouds were a dark indigo, but the sun itself wasn't visible yet.

Two police officers were standing by the kerb, arms stretched out to their sides, shepherding the crowd. I recognised their faces but I didn't know their names.

'Don't stay on the roads. Go to work, go to school, everyone.'

'Is it true?' asked our neighbour, Old Li, who was standing next to me. 'Eight per cent?'

'If that's what they said,' one of the officers said.

'Do you have a plan?'

'We're police officers. We just have to make sure nobody panics and gets hurt.'

'But that's not a plan, is it?' I heard Old Li's wife whisper to her husband.

I worried about Dong Ji. She'd always been much more affected by news of the sun vanishing. Every time it happened, she'd tell me that she was going to leave Five Poems Lake, though she hadn't made any attempts to act on her words. But eight per cent was a shock, even to me, so I couldn't rule out the possibility that she may finally pack her bags and venture into the desert.

My first priority, I told myself, was to make sure the pharmacy would continue to run without much disturbance. It was precisely during times like these that people needed medicine the most. I went back inside and phoned my supplier.

'Is Driver Hua on his way?' I asked.

'He should be there soon.'

'Good. Just wanted to make sure the delivery is still coming.'

'I still have to make a living, don't I?'

'Good.'

He hung up. I tried calling Dong Ji next, but nobody answered. It must've been too early. I went into the shop, grabbed my rolling paper and tobacco, and walked out from the back, leaving the door wedged open behind me. I made myself a cigarette and smoked while I waited.

The truck arrived before I could finish.

'You told me last time that you were going to quit,' Driver Hua said as he opened the trailer doors. 'The sun got you worried?'

'Are you not?'

'You know my wife said that she has plans for after the sun disappears completely?'

'How is she so sure that will happen?'

He looked at me as though the answer was obvious.

'We lost ten per cent of the thing today,' he said.

'Eight.' I released my cigarette into the road drain. 'Eight per cent. That's what they said, anyway.'

He jumped into the back of the truck and started moving boxes onto the tailgate.

'Is this everything?' I asked.

'Most of it, yeah. Boss said he'll contact you directly about refunding for the missing stuff.'

I went to fetch the metal trolley from the storage room as Driver Hua began to unload the boxes.

'So what are your wife's plans?' I asked.

'She told me a few of her ideas, but the way I see it, none of it is going to make a difference. Plans don't matter when you're dead.'

When all the boxes were in the storage room, I handed him some cash and thanked him. I could see the sun now. It was not a half sphere any more and looked like a crescent moon splashed red. The new curved side gave it a timid quality, a modesty that I'd never associated with the sun.

'Wait,' Driver Hua said, right as I was about to go inside. 'How's Miss Pan doing?'

'Look, go back to your wife and stop worrying about other women.'

'I just want to know how she's doing.'

His voice became louder, more sure, showing that there was nothing for him to hide.

'I'm not committing a crime here,' he said. 'And I don't worry about other women. Only her.'

'Her belly's growing pretty big.' With my arms, I gave him an idea of how pregnant she was. 'Out to here, I'd say.'

'You still don't know who the father is?'

'I don't think anybody knows.'

He lowered his voice and bent down a little to match my height. He smelt like a soggy cigarette butt. 'Do you think you can ask her?'

'I told you last time. It's none of my business.'

He looked down and began fidgeting with the money in his hands.

'Can you do me a favour?' I asked. 'I have something I need to deliver to her before the end of today, but I can't leave the shop unattended.'

His head shot up and he nodded enthusiastically.

'Wait here,' I told him.

I fetched the container of bird's nests out of the fridge and wrapped it in a linen towel before bagging it. Back outside, I handed it to Driver Hua and warned him that it was expensive.

'I won't spill a drop,' he promised.

'Leave it with the security guard,' I said. 'And don't say anything inappropriate.'

He laughed. 'What could I possibly say? I've delivered things to her before. Has anyone complained about me being inappropriate? I'll head home and take a shower before I go.'

He climbed into his truck and secured the bird's nests in between his legs and up against his crotch. By the time Driver Hua drove off, what was left of the sun had risen high.

I called Dong Ji again. The receptionist at the wellness parlour told me that she was already in the middle of a treatment. Relieved that she was at work, I asked the receptionist to tell her to call me back during her break.

The Su girl was waiting outside once again when I

opened the shop. She must've taken a liking to visiting me. She was wearing the same thing as the day before and the little white scarf was tied into a bow in front of her neck. Under her arm, she was holding a basketball.

'Come on in,' I said, glad to have some company to distract me from the news of the sun. 'You barely have any fat on you to keep you warm.'

Miss Pan's fur gloves were still in my lab coat pocket. I handed them to her.

'Put these on for now,' I said. 'Your hands must be freezing.'

She thanked me but shook her head. Then she spent some time fiddling with her scarf, adjusting the bow so that it would be centred correctly.

'Shouldn't you be at school?' I asked.

'I don't go to school, but I should be at work.'

'Aren't you too young to work?'

'I'm eighteen!'

'I don't know if I believe you.'

'It's your word against mine,' she said with an exaggerated shrug.

'So why aren't you at work?'

'I did go to work,' she said. 'I'm a seamstress. I can't do much else with my bad leg. This morning, I was hemming some trousers and then, out of nowhere, I just got the desire to walk. To move my body.'

She stroked the basketball.

'I found this by the court on my way here so I picked

it up,' she said. 'I don't even know why. I've never played basketball before. I can't really play with my bad leg. You can imagine.'

'You can shoot baskets,' I said. 'You don't have to move your legs too much. My sister and I used to do that.'

She gently tossed the ball into the air with both of her hands and caught it again.

'I need to buy some medicine for my brother,' she said, as though suddenly remembering why she'd come.

'What's wrong with him?' I asked.

'He can't sleep. He has nightmares all the time.'

'How long has it been?'

She shrugged.

'All his life, I guess. But it's been especially bad in the past few days.'

'I don't blame him.'

I slid open the glass door of the display case and took out a box of jujube seed pellets.

'Here,' I said. 'Try this. It should calm the mind and help him sleep more soundly. If it doesn't work, then I suggest you take him to see a doctor.'

We chatted for a little while longer until she paid for the medicine and said she had to go. A few customers came in during that time. They all knew exactly what they wanted, so they were in and out quickly. That was one of the perks of owning a medicine shop. Nobody came in without a purpose. Most people never lingered.

'Are you going back to work?' I asked the Su girl.

'I'm going to the Western District.'

'Why?'

'To get a haircut.'

I didn't ask her why she was going all the way across the lake for a haircut, but she must've read my mind.

'There's this fancy hair salon there,' she said. 'Mirrors everywhere. High ceilings. Everything made out of wood. I've always had a desire to try it out. Splurge a little bit. Call it a compulsion. Just like how I wanted to leave work, move my body, pick up a basketball that I'm not going to use. Makes sense, right?'

'I guess so. As long as it makes sense to you.'

'I knew you'd understand.'

Right before she walked out the door, she turned towards me and held the basketball in the air, pretending like she was going to shoot.

Dong Ji returned my call during her lunch break. I'd completely lost track of time; the pharmacy had grown overwhelmingly busy with customers dropping in to buy all sorts of emergency medications. I'd been in the middle of packing some huang lian for a customer when the phone rang.

'What's wrong?' Dong Ji asked. 'Sorry, it's been one client after another this morning. I guess lots of people need relaxing.'

I could hear another woman speaking quietly in the

background, though the exact content was difficult to make out.

'It's been busy here too,' I said.

'So why did you call? Are you all right?'

'Everything is fine,' I said. 'I just wanted to make sure you're OK. The weather forecast must've been, I don't know, disappointing to hear, I guess.'

'Disappointing?' She switched to a whisper. 'It's horrible for all of us, isn't it? Are you staying warm? I have an extra coat in the dorm—'

'I'm fine, Dong Ji.'

I bound the pack of huang lian with string and scribbled down the price on the brown wrapping paper. The woman handed me the money and thanked me before turning away.

'Remember to keep all the ginseng for yourself,' Dong Ji said. 'You're going to need it to keep your body warm.'

'I've saved a few for you.'

'Are you sick? Your nose sounds stuffed up.'

'No,' I said, trying my best to speak in my usual voice. 'I'm not sick. I'll come drop off the ginseng I kept for you.'

'I'm fine. Just eat it yourself.'

'It's not like you don't have to stay warm too.'

'I have a stronger body,' she said. 'I have more fat on me. Plus, they keep the temperature warm here. Are you sure you didn't catch a cold? Your voice—'

'I feel fine,' I said. 'I'm not sick.'

The next customer, an old and wrinkled man, approached the counter and asked for patches for joint pain.

'You're not leaving, right?' I asked hesitantly into the phone.

I heard Dong Ji inhale on the other end of the line, which was followed by silence. She had the habit of holding her breath when she concentrated on something. I grabbed the patches for the old man.

'I wanted to call you this morning and tell you we should leave,' Dong Ji said after a sigh. 'I really did. I'm so tired of this place. But there are things I have to do first. I can't just get up and go. I'm not my mother.'

Reassured by her words, I told her to get back to her work and that we'd see each other soon. She reminded me, one more time, not to sell the ginseng to anybody.

'Take some cold medicine,' she said before hanging up the phone. 'I still think your voice sounds strange.'

Dong Ji and I had different mothers. Hers left when she was eight years old. Everyone assumed that she'd fled Five Poems Lake, because she hadn't been seen anywhere since then. Even the police couldn't find her. No one knew how far she ventured, let alone whether or not she made it to another place. As for my mother: I don't even know who she is. It seems that Ba was particularly attracted to the allure of mysterious women. He brought me home one day in his arms, declared that I was his daughter, and that was all there was to it. Sometimes, I even question whether my father knew who bore his child. When I was still young and had a more colourful imagination, I'd wonder whether I was the baby of a woman he'd saved from a criminal. Maybe he had a fleeting romantic relationship with her. I also considered that he could've found me all by myself, abandoned in front of the police station. Thinking through these scenarios didn't upset me. They were a child's curiosity more than anything. I really had little interest in the truth. I did not need more than the people I already had.

It never made much sense to me why Dong Ji had chosen to work at a wellness parlour. Growing up with her, I knew that all she wanted to do was leave Five Poems Lake. It hadn't been in her plans to take over the pharmacy, but Yeye died before I came of age, so she had to manage the shop while supporting me as I finished school. I told her I could drop out, but she refused to listen, even scolded me for offering.

She never explained to me why she despised Five Poems Lake so much, but it wasn't difficult to figure out that it was because of her mother. Dong Ji, unlike me, who couldn't care less about a mother I'd never met, had always been chasing after hers. It wasn't that she wanted to find her – if that was her goal, she would've acted upon it long ago. I think it was a yearning to be closer to her mother, in mind, in heart, in spirit, in ways that were more complex. Dong Ji wanted to leave, not to go where her mother was, but to do something that her mother had done. Maybe that required more courage.

It was not in Dong Ji's nature to tell me any of this outright. On occasion, when she did say something about her mother, I felt guilty, as though I was more fortunate because I'd never known mine. Time and time again, the pain of loss has proven itself to be deeper, more unabating, than the pain of not ever having. I didn't need anyone to tell me this.

The first thing I did when I finished school was tell Dong Ji to go and live her own life. It still astonishes me to think that I'd once been so sure of myself, that my youth had given me the reckless bravery needed to let her go. She had been reluctant, saying that we could run the shop together, but I reminded her of her promise to leave the business to me once I was old enough. Knowing how stubborn I could be, and maybe feeling a slight sense of relief, she finally conceded.

A few days later, she told me that she'd enrolled in a course to become a beauty and massage specialist.

'Where?' I asked.

'At the wellness parlour that Miss Pan's mother owns. West of the lake.'

'Are you happy about it?'

'It offers a lot of security,' she said, which surprised

me. I hadn't expected security to be the thing she wanted.

'What about you?' she asked. 'I'm worried about you managing the shop all by yourself.'

'Oh, I'll be all right.'

'What if the suppliers cheat you?'

'Stop worrying.'

'I'm not worrying. Fine. I won't worry.'

'Are you going to commute from home?' I asked.

'They have somewhere for me to stay.'

'That's nice.'

'Yeah, it is.'

She moved into the parlour staff's dorm room with a bag of her prettiest summer clothes, which she'd only worn for one year before the weather became too cold.

Miss Pan came in the following afternoon to buy dried longans. There weren't any other customers and I was glad that we were able to chat. This time, she was in a cream-coloured padded coat that wrapped around her like a puffy duvet. It was impressive how the silliest things could look charming on her.

'I'm moving into the hospital today,' she said. 'All by myself.'

She made a comically sad face, which then turned into a playful smile.

'And you call yourself a socialite?' I teased.

'Please, nobody calls themselves that.'

She fell silent for a moment, and though it was a natural pause, I thought I sensed an anxiety. Like a photograph that was out of focus, I saw a vague image of her, blurring into a colour, losing her edges.

'I think it's really brave,' I said. 'To have a child.'

'Someone's got to keep us going, right?'

'I can come with you,' I said, rather impulsively. 'I'll just finish things here a little earlier today.'

Her pregnancy had always felt close to me. I'd been seeing her regularly – every week she came by to pick up bird's nests – but that wasn't the only reason. It was because she reminded me of Ba. Maybe it was the secret other half. That her child only had one parent. I'd often wondered whether Miss Pan planned on telling her baby who the father was, but I couldn't bring myself to ask.

'That would be so lovely,' she said.

Miss Pan took her time articulating those words, so that they seemed to hold more weight than usual.

'I have to go get my things,' she said. 'But I'll come pick you up in two hours. Does that give you enough time?'

'Tell me which hospital and I can meet you there,' I said. 'You don't have to pick me up.'

'Nonsense.'

Not long after she walked out the door with her longans, Driver Hua entered with a beer bottle dangling from his hand. A few days a week, he'd go drinking with his friends at The Old Sister, a small dumpling restaurant in our neighbourhood. Sometimes, he'd stop by my shop on his way home. I suspected it was because he wanted to see Miss Pan, but never once had he been fortunate enough to catch her. Today was the closest he'd ever been.

'You just missed her,' I told him. 'You're not driving your truck like this, are you?'

'I didn't miss her,' he said, burping loudly. 'I saw her leave.'

He was struggling to keep his eyes open. With each extended blink, he looked like he was going to fall asleep standing up.

'What was she here for?' he asked.

'Why shouldn't she be here? She's due soon. I'm going to the hospital with her.'

'Why does she bother coming all the way here when she could just send somebody?'

'Maybe she likes talking to me.'

'You think people like her enjoy talking to people like us?'

'I'm not like you. Look at yourself. Where's your truck?'

'Truck, truck, truck. Is that all I am to all of you? A truck driver? I'm not allowed to just come and chat? You think you're better than me too?' He tried to burp again, but nothing came out.

'Go home, Driver Hua. You're drunk.'

'My name is Hua Ge,' he said. 'Remember that. Which hospital is she going to?'

'I can't tell you that.'

'You think I can't figure it out myself? People like her only ever go to New Hope.'

'So?' I asked. 'What are you going to do with this information?'

He finished his beer in one gulp.

'You shouldn't have let me deliver that stuff to her,' he mumbled.

He turned around and hauled himself out of the shop, his bottle swinging like a pendulum.

'Don't throw that in front of my shop!' I shouted after him.

Most of the time when he stopped by, he'd only be slightly tipsy. The only other instance he came in this drunk had been a week ago. He'd spent an entire hour foul-mouthing his wife. She had to be cheating on him, he told me, because he had called home while he was at work and she was breathing heavily. She must've been naked in bed with another man, going at it, while speaking on the phone with her husband. He was disgusted at how shameless she'd become. He called her a whore, among other things.

I didn't think any of his suspicions were reasonable, but I didn't know the entire story and frankly, it had nothing to do with me, so I didn't say anything. It felt wrong to meddle in other people's private affairs. Driver Hua used these accusations to justify his infatuation with Miss Pan. I couldn't quite understand how he could let his emotions get the better of him, and then dump them on another woman. I wondered whether it really made him feel any relief.

Despite all that, like I said, it wasn't up to me to judge whether he was a good or bad man. All I could say was that he was a good truck driver. Punctual and careful. Never drunk on the job. That was all that mattered to me, anyway. The day after he complained about his wife to me, he even came all the way here to apologise. It seemed he really did feel sorry, but within a few sentences of his apology I could tell that he didn't have a clue what he was sorry for. Watching him, all that stirred inside me was a feeling of pity.

New Hope Hospital was located in the Southern District. It sat on a small bit of land that jutted out onto the lake. Of the two hospitals, New Hope was the better one, although the exterior was dispiriting, painted in black and grey stripes, with each shade representing a different floor. The paint was flaking off the walls and there were large water stains on the bricks. Even the hospital couldn't avoid falling to pieces.

Five Poems Lake was a rather monochromatic place. If you stood on the tallest building and gazed down at the town, you would've wondered why it was all so grey, like everything had been drawn in pencil. Of course, I am not claiming that there were no colours in the town – there were plenty, in fact – but when viewed from afar, all those colours would somehow blend into the same leaden hue. I couldn't remember if it'd always been this way, or if the town had gradually forgotten its colours in the past twelve years, as trees darkened to black and grass turned to dust.

The maternity ward was on the top floor and, unsurprisingly, it was mostly empty. Looking out from the windows, the location of the building made the rooms feel like they were floating. Good for the emotional wellbeing of new mothers, I imagined, so long as they stayed inside. Miss Pan had a room to herself. Beige walls. White duvet. Pink curtains. Everything in the room was in one of those three colours. Miss Pan fitted right in with her cream-coloured coat. It was as though she was in a painting. Things in real life never usually matched up quite so nicely.

There was a television mounted on the wall over a dresser. I helped her put away her belongings: clothes, bottles of skin products, make-up, a brand-new diary, baby girl clothes, massage oils, a camera. She treated me to an impressive plate of shredded pork in sweet bean sauce, completed with a stack of spring onion pancakes that she'd asked her driver to pick up for us. We ate in her room and talked about all sorts of small and pointless subjects. Neither of us mentioned the sun. I had the feeling that she only wanted to speak about hopeful things. Maybe it was because of the baby. They all say that the foetus can hear inside the womb.

I didn't leave until around ten in the evening.

'I won't be able to come tomorrow,' I told her. 'I have to watch the shop. But that doesn't mean you should be alone. What I'm trying to say is that you need to get someone to take care of you, Princess Pan.'

'I'll find someone,' she said. 'I'll call the maid.'

I made her promise before I left. She accompanied me to the staircase. Watching her walk back to her room in the quiet hospital, it occurred to me that nobody can be entirely spared from solitude, not even Miss Pan.

She'd offered to have her driver take me back, but I wanted to ride the bus. Few things comforted me more than a running bus late into the night. I reckon that anyone who lives alone must experience nights when loneliness nags at them, keeps them from falling asleep. When that happened to me, I'd take the bus all the way to the Western District. The soft night lights coming in through those large glass windows always succeeded in calming me.

The bus stop was just in front of the hospital and I had to wait almost half an hour in the cold before the bus finally showed up. The driver didn't even bother to check the time and immediately drove off after I paid the fare. At moments like these, I saw that bus drivers were in positions of power. Only they could decide when to arrive and when to leave. Everyone else's plans had to be adjusted accordingly.

There were three other people on the bus. A couple was sitting towards the back, and a man in a long coat was standing by the door. Though it was only twenty minutes away by bus, I hardly ever had reason to venture to this southern part of the town. The University was located here. On a weekday like this one, the streets were quiet and dark at night, lit only by the fluorescent convenience store signs. The only people out were students buying their late-night snacks and drivers taking a break, smoking by their trucks. The bus motored down the road, passing all the empty stops without slowing down. The night felt brief and long, close and far, warm and cold, all at the same time.

Back in the east, there were only the faint sounds of the wind. I checked my watch. It was almost midnight. It'd been a few hours since dinner, but I could still feel the oil and sugar from the shredded pork coating my mouth and churning in my stomach. I felt greasy, which really made me want to take a bath and drink some tea. Dong Ji had told me about the baths they offered at the parlour. Dried flower petals, herbs, salts and honey masks. They would even scrub your back for you. Maybe I should save up for a treatment, I thought. I should do it before the sun disappeared, because who knew what the world would look like once it all turned to darkness.

Whenever Dong Ji visited me, which was normally every Thursday, she would tell me all about her work and the strange things she had to do, like changing the intonation of her voice to fit the music in the background or the level of light in the room. One of the most expensive treatments involved smearing honey onto naked customers. With the bees gone, honey was impossible to find at the market.

It was rather dark when I got off the bus. The convenience store next to the pharmacy usually kept its sign on throughout the night but, with the sun fading so much all of a sudden, maybe they wanted to save on electricity bills. As I walked, I imagined myself paying my visit to Dong Ji's workplace. It was strange to think that I'd never been inside. I wanted a honey mask, I decided, like how the Su girl had wanted a haircut at that fancy salon. I was happily thinking about myself naked, silken with honey and soaking in hot water, when I turned into the alley and saw Driver Hua standing in front of my back door. They hadn't put up any streetlights – the walls were too close to each other – but his face was clear under the blue moonlight.

We both froze. I'd always had plans for a situation like this one, where I'd find myself alone with a man on a deserted street at night. All of them involved some effort to escape, so I couldn't be sure why I just stood there. Maybe I was afraid, but my heart wasn't pounding and I didn't have the slightest urge to scream.

Fear didn't manifest in the ways I'd always imagined it would.

I watched him and he stared back at me. It was just Driver Hua, a man I saw all the time. He wouldn't do anything to me.

He took a few steps forward and stood in front of me. Was I asleep? I was sure that he had been farther away. Only in a dream would he be able to get to me so quickly from that distance.

'I tried to give up,' he said, breathing heavily. His eyes were bloodshot and he reeked of booze.

'I tried,' he said. 'But I can't stop thinking about her. She's a swan in the sky, and I'm down here, in a pile of shit and mud. I can't stop imagining just how pure and clean it is up there. And you brought me so close to her, and now I'm more aware than ever that her life and mine will only ever run in parallel. How could you do that to me?'

He paused and pressed his forehead against mine, his eyes glaring at me. I shoved him and he stumbled backwards, almost toppling over before regaining his balance.

'The police station is right there,' I warned him. 'Don't touch me.'

He stood in between me and the back door to my house. If he wanted to stop me, there was no way I could push past him. The alleyway was too narrow.

'What do I do with this burning desire?' he asked,

like he really expected me to have an answer. 'You think if I just tried hard enough to win her love, she'd want to fuck me in my truck? I'm too afraid to even look at her.'

He began to laugh and cry at the same time.

'I don't even know what she smells like,' he added.

'You'll only get yourself in trouble if you keep obsessing over her,' I said, trying to calm him down. 'You don't know her. She's just . . . an idea to you. You told me once that you like quiet women. She's many things, but certainly not quiet.'

I tried to force a smile, but instead I felt my facial muscles twist into a stiff and dishonest expression.

'Have sex with me,' he said. 'Please. You're the closest I'll ever get to her.'

'You've lost your mind,' I said, stunned by his suggestion. 'I want nothing to do with this.'

'Have sex with me,' he repeated. 'Just once. I need to feel her. Just this one time.'

His insistence disgusted me and I stormed towards my house. As I pushed him aside, he grabbed onto my wrist and sniffed my hair like a dog. The smell of wet tobacco and beer invaded my nose. Combined with the greasiness in my mouth, it almost made me throw up. In panic, I shook my arm with all my strength, surprised to find that his hand slid straight off.

When I turned my head to make sure he wasn't following me, I saw him standing with his back curved like a shrimp next to a pile of garbage bags, fumbling around his crotch and unzipping his trousers. I tripped over a loose brick and almost fell. I heard his loud breathing quickly turn into a series of moans and felt a heat on my back, like I was standing near a flame. When I spun around, Driver Hua's trousers were halfway down his legs, and for an instant, I saw the outline of his face before a bright light pushed itself out of his open mouth, flared up and engulfed his head, blinding me.

I don't know how much time it took for all that to happen. It must've been a matter of seconds. Coming back to my senses, I dashed for the door, though my eyes were wincing from the light. All I could see was a white screen over everything. Or was it black? I battled with the lock and pushed my way in. I wheezed and coughed and spat out the saliva that'd been bubbling in my mouth. I climbed onto the back of the sofa and pulled the curtain aside. As the whiteness blurring my vision cleared away, I saw Driver Hua standing, straight as a pole now, hands by his sides. Up where his head should've been was a radiating light that illuminated the alley like it was daytime.

A few more seconds passed. Or maybe it was a few minutes. He began walking to his truck, which was in the same place he usually parked when he made his deliveries. He moved with purpose, like he knew exactly where he was going, but with no sense of urgency. His trousers had slid to his feet and almost tripped him as he tried to climb up the steps to the door of the truck.

After some struggling, they fell off entirely and he pushed himself into the driver's seat. The light shined atop his shoulders, beams bursting from where his face had been.

After he started the engine and drove away, I waited a while to make sure he really was gone and rushed into the kitchen and ran my hands under the tap. I was sweating all over. I started washing my neck with the freezing water, which sent a searing jolt through my body. The grease from the pork shot up my throat. I heaved it all out into the garbage bin. I stared at the dirt underneath my fingernails, feeling filthy.

It was like a small sun, I thought. His head was a small sun.

2

The first thing Dong Ji did when I opened the door to her was stick a cigarette in between my lips. I recognised it as one of our father's cigarettes, the kind he would sometimes buy as a special gift for himself. They were slender cylinders assembled by machines, not the kind we rolled ourselves, and so there was more tobacco packed into each one. There weren't many of those left when Ba was alive and the few machines that made them had all broken down since then. Ba had left behind an unopened pack of twenty in his drawer and Dong Ji had been keeping it safe with her. Before now, we'd only ever smoked two cigarettes out of the pack, on the day Dong Ji moved out and I took over the pharmacy.

'That perverted bastard,' Dong Ji said, rubbing my arms as though trying to warm me up. 'Are you hurt?'

'He was by the back door,' I said. 'But he drove off.'

'I've been telling you to go through the shop entrance, especially at night. How many times have I said that?

How come you never listen to me? That back alley is too dark. Have you gone to the police?'

With the lighter she had in her pocket, she lit the cigarette for me and then for herself.

'I don't want to go outside,' I said.

I followed Dong Ji to the living area. She tossed her coat onto the dining table and led me to the sofa. She was in her boat sweater that she only ever wore at home. It was green with a white boat knitted in the front, the boat that belonged to the Su family. The sweater was so old that the green had been washed into a pale pear colour.

'Has he lost his mind?' she said. 'To think that Driver Hua would dare to do something like that right next to the police station. This never would've happened if Ba was still alive. We have to report him tomorrow.'

Dong Ji sat leaning forward with her elbows on her knees, her hands interlocked, one of her legs bouncing up and down. I mounted the back of the sofa, pulled open the curtains, and slid the window up a little. The cold breeze, as though it'd been waiting, surged in through the gap.

Outside, only moonlight shone on the pavement. I found it strange that the moon was still so bright, given how little sun there was left.

'Let's tell Gao Shuang,' Dong Ji said. 'We should know where Driver Hua lives. Yeye should have it in his address book. He is not getting away with this. Makes me sick.'

'All right.'

She looked at me. 'What is it?'

'What do you mean?' I looked down at the yellow filter of the cigarette.

'It seems like I'm more troubled by this whole thing than you are.'

I reached my hand out the window and stubbed out my cigarette on the wall. I tried my best to finish it as I did not want to waste something so precious, but the taste of it reminded me of Driver Hua. By the last few puffs, I just didn't want anything to do with it any more.

'Give it here,' Dong Ji said. 'Don't throw it on the pavement.'

She took the cigarette from me and went into the toilet. I heard her flush.

'While he was doing it,' I said when she returned, 'his head was eaten by a sun.'

Dong Ji stood by the sofa and blinked at me, so perplexed that her face couldn't even find the right expression. I repeated what I'd said.

'What on earth are you going on about?' she finally asked. 'Are you feeling all right? What does that even mean?'

'I don't know how to explain this any other way,' I said. 'A light came out of his mouth, and the next moment, I couldn't see his face any more.'

'Stop joking around with me. This is not the time. Do you even realise how much danger you were in?'

'Listen to me.' I looked at her dead in the eye. 'I don't understand either, but that's how it happened.'

Upon seeing that I was serious, I saw her worry gradually shift away from me and towards what I was saying. It must've sounded like an impossible series of events, but she knew that I'd never been one to make up or exaggerate things.

'A light, you said?' she asked.

'Yeah. It was like a sun. Hot and bright.'

'He had a sun on his head?'

'Not on his head. The sun *was* his head.'

'The sun was his head . . .' she repeated under her breath.

She plopped down, leaned back, and then forward, and when she rocked back again she turned her body around to look out the window.

'Was it a full sun?' she asked.

'It was so bright. I couldn't see.'

She pulled her legs onto the sofa and the extra weight made her sink deeper into the cushion.

'You're in shock,' she said. 'Imagining things.'

'I know what I saw.'

She buried her chin between her knees.

'Do you think the police would believe me?' I asked.

'I don't even believe you.'

'Maybe Gao Shuang would.'

'Maybe. Though I doubt it.'

Despite what I'd said, I hardly believed myself. Eyes

played tricks on the mind all the time and, like Dong Ji said, I was in a state of shock. Ba had once told me about a kidnapping victim who couldn't remember the face of the perpetrator, even though she'd been confined in the same room with him for an entire day. He'd called it dissociative amnesia.

'The police need to know about what Driver Hua tried to do,' Dong Ji said. 'It's up to you whether you also want to tell them about this sun thing. I wouldn't, if I were you.'

She waited a moment and then sighed.

'I miss Ba,' she said.

'I miss Yeye.'

'Me too.'

It was lung disease that took Yeye. Those two packs a day for fifty years caught up with him in the end. A few days after he was admitted into the hospital, he told me that he did not regret having a single one of those cigarettes. Without them, he said, he wouldn't even have made it this far. Then, he laughed a little, which made him cough and hurt him somewhere, everywhere, but still he laughed some more. Because of that, my memory of that time was made more bearable.

In Yeye's final weeks, he forgot about us. The only person he recognised was his nurse. His eyes became cloudy, brushed over with a layer of something watery, like he was always on the verge of crying. All things considered, he seemed happier that way, enshrouded in fog, remembering only wisps of his life. Seeing him like that brought us some relief. We were told that eighty-one years was a long time to live, and knowing this, we should have felt comfort, too, but what I saw in his long life was the sadness of all the things that had passed for him.

I'd imagined that the day he lost his memory was the real moment that death came for him, but as I watched the sun disappear over the years, I began to wonder whether this erasure was, in fact, a form of death for all of us. Who are we if not just a collection of memories locked in a container made of flesh and blood? If we ceased to exist in the memory of those closest to us, then how could we assume that we were as alive as we'd been before? It made me realise that, despite what people said about it, death was not so straightforward after all.

Dong Ji had to go to work in the morning, but she said she'd take the afternoon off and watch the store so that I could go to the police station. Before she left, she handed me a box of medicine and a cup of hot water.

'Here,' she said. 'I heard you coughing at night.'

I'd been feeling better, but the cigarette had irritated my throat again. I took the medicine from her.

'Listen to your older sister,' she said as she put on her coat. 'You must tell Gao Shuang.'

I nodded.

'Will you be all right?' she asked. 'I can call in sick.'

'I'll be fine.'

After she went to work, I had some time before I had to open the shop, so I headed to the convenience store and bought a copy of the newspaper. There was a chance that Driver Hua was in there. Maybe somebody else had seen him. If he'd gone home, his wife might've reported the incident.

I avoided the back alley and opted for the longer route. It was still dark outside. Whenever the image of Driver Hua's face crept into my mind, I tried to wipe it away and focus on what was in front of me, wondering how thick the ice over the lake had become. When the temperatures first dropped enough to freeze the water, I often saw people sledding on the ice. They built wooden chairs with sleds instead of legs, which they sat on to glide around the surface of the lake. These activities went on for a few months until the sun faded again and it became too cold.

I made it to the store, purchased a copy of the newspaper and carried it back home with me before realising that I didn't want to read it. I stashed it away and turned on the television. The weatherman was on again. He was wearing a solid orange tie.

'Today is going to be cloudy with a chance of snow in the evening. The sun has remained the same, but we are beginning to feel the effects of the eight per cent fading. Days are going to get much colder, so residents should stay warm . . .'

I thought about calling Driver Hua's boss to see if he

had shown up to work but decided that I'd wait until the afternoon. I have to admit that I was afraid to tell Gao Shuang. I couldn't predict what he would do to Driver Hua: he'd always had a temper, treating me like a little sister that he had to protect, especially since Ba died.

I was boiling some eggs for breakfast when the phone rang. I had no appetite, but Dong Ji had made me promise to eat something. Some food would do me good, she said, calm my nerves. She was always right about these things, so I listened. Plus, the bitter taste of the medicine had refused to go away.

One of the eggs cracked when I dropped it into the boiling water, and the yolk and the white burst out of the shell and morphed into a pale yellow, monstrous mass. Sickened by the sight, I moved the pot off the heat and left the eggs in the hot water.

When I picked up the phone, a man with a deep voice introduced himself as the Director of New Hope Hospital. I remembered that when Miss Pan checked in, I'd left my phone number with the receptionist as her emergency contact.

In a rehearsed and respectful tone, he told me he was very sorry, and that he was prepared to do anything to help. After going over some other things I didn't understand, he paused and then told me that sometime in the middle of the night, Miss Pan had jumped from her window.

*

Are those who choose to end their own lives too cowardly to live, or too fearless in the face of death? Do they do it out of strength or weakness? I reckon that since such judgements are only levied by those who live on, they are, by nature, irrelevant. Perhaps ending one's own life is only a solution: neither good nor bad, like a mathematical equation.

I spent a long time sitting by the phone, turning these questions over in my mind. I couldn't believe what I'd heard. Miss Pan was different from all of us, but she also seemed like us in so many ways. There were plenty of things about her that would have surprised me, though none were beyond the realm of my imagination. Yet, no matter how much I thought about it, I couldn't imagine her dead. Her presence had always been like a bright light in a dark room. Everywhere she went, it was as though air itself would part for her. She'd carved out a space of her own in this world, and it was impossible to think that she no longer occupied it.

She was naked when they found her. The Director told me that she was gone 'with her baby'. My first impulse was to think that Driver Hua must've got to her, but I was told that there were no signs of someone else being involved. Though it really did not make a difference, I still asked whether she had delivered the baby before she died, to which the Director responded, 'Thankfully, no.'

While Driver Hua's head was being gobbled up by the light, Miss Pan was falling to her death. I began to resent Driver Hua for this coincidence, even more than for what he did to me. I found it horrid, the idea that he could've occupied, even for a second, the same space between life and death, human and inhuman, pain and relief as did Miss Pan, who, with her unborn child and her naked body, was so untainted and beautiful and ready to be a mother.

I couldn't bring myself to think about the baby. Every time my thoughts steered in that direction, I would see, in the far distance of my mind, a dark space, a womb turning cold, and that image terrified me to no end.

I tried to find something to preoccupy myself with and I remembered the eggs I'd boiled. In the kitchen, seeing the cracked egg floating in the lukewarm water, I thought I smelt something rotten, and I dumped all of it into the garbage.

When Dong Ji returned from work, she had already heard about Miss Pan. News travelled quickly. Everybody at the parlour had been talking about it. When Dong Ji took my lab coat from me, the fur gloves were still bulging from the pocket. I didn't have to tell her where I'd got them from. Even when Dong Ji was in charge of the shop, Miss Pan would give her nice things too from time to time.

Dong Ji handed them to me. 'Don't lose them,' she warned.

The fur was soft and delicate. They reminded me of

Miss Pan's hands – evidence that she'd never done a day of labour in her life. I'd once thought that childbirth was going to be the most strenuous thing that she'd ever have to go through. How wrong I'd been.

Now that Dong Ji had come to watch the shop, I was free to go to the police station. Just as the weatherman had predicted, the sun was shining a soft, cold light behind a thick layer of clouds. It was early in the afternoon, but the sky had already dimmed, as though night could come at any time. They'd even turned the streetlights on.

How much had Miss Pan's world failed her, if death was the only way for her to communicate how she really felt? Just how ignorant was I? I was turning these thoughts over in my mind as I walked past the metal gates and into the police station. I hiked up the stairs to the fourth floor and walked to the left end of the hallway. Two officers who knew my father greeted me on the way. The worn walls trapped the scents of cement, sweat and cigarette smoke.

Every time I came to the police station, I would see traces of Ba everywhere, in the chairs, desks, the brass door handles, the medals in the display cabinets. I felt his presence here more than at home. If ghosts really existed, then his must've lingered in the police station.

I poked my head into the office where Gao Shuang worked. There were three men inside, but Gao Shuang wasn't there. I only knew one of them, the

Su girl's brother, Big Su. Big Su was, in fact, quite small and reminded me, oddly, of a fire hydrant. His limbs were short even for his height, but he had stocky shoulders and a rather big head. There was a large birthmark on his left cheek. The fatigue was clear on his face. When I entered the room, he was chatting with a man who was in casual clothes. The third was eating roasted chestnuts at his desk while reading the newspaper.

Gao Shuang served under Ba for two years after he graduated from police college. He told me that, in that brief amount of time, Ba taught him everything he needed to know about being a police officer. During the months after Ba died, he came by every day with bags of groceries in his hands and, on the more sentimental days, tears in his eyes.

I heard somebody call out to me and turned around to see Gao Shuang jogging down the hallway.

'Oh,' I said. 'There you are.'

'"Oh, there you are."' He made an unenthusiastic face that was supposed to resemble mine. 'Would it kill you to look like you're excited to see me?'

'Sorry.' I feigned a smile. 'Can I talk to you about something?'

'Yeah. Of course. What is it? Come on in.'

Big Su and the man in casual clothes paused their conversation and nodded at us. The man in casual clothes patted Gao Shuang on the back as we walked by. The officer eating chestnuts also looked up from his paper and waved at me. Gao Shuang guided me to his desk and pulled a second chair over. We sat facing each other. I could sense all three men assessing me with their eyes. Their gaze made me nervous after what had happened the night before, so I tried my best to avoid their eyes and look only at Gao Shuang.

'I heard about Pan Yong,' Gao Shuang said softly. 'I just came back from the south side. I went to see a guy I know at the station. Pan Yong was one of your customers, right?'

I nodded.

'Did you know her well?'

'I was at the hospital with her yesterday.'

An expression of concern rose to his face.

'Did you—'

'No. I left at around ten in the evening.'

'I'm sorry,' he said.

He started squeezing his left thumb with his right hand. He often did this when he struggled with finding something to say.

'Was it a murder?' asked Big Su.

I looked over. The hair on the back of his head had been flattened by his pillow. When he spoke, he had the side of his face with the birthmark slanted slightly away from us.

'I don't think so.' Gao Shuang paused with the thumb. 'But they don't have all the information yet, so they can't rule it out completely.'

'Wasn't she pregnant?' Big Su asked. 'Who was the father?'

'She never told anyone,' I said. 'My sister said that not even her mother knows.'

'Maybe it was the father of the baby,' he said. 'The killer.'

'You're a police officer,' Gao Shuang said sternly. 'Don't go around making groundless claims.'

'How about you stop telling me how to be a police officer?' His voice became louder, more aggressive.

The man at the desk poked his head up and shouted, 'Can you two keep it down? I'm trying to read the paper.'

'You're the one who is shouting,' Big Su said. 'Look, maybe it was suicide, but that just goes to show how weak-minded these rich people are. There's no reason for her to be discontented. On top of that, she killed her own baby. Her own child. It's murder. It's evil. Sure, maybe her mother didn't love her, or maybe she fucked too many men and she felt guilty—'

Gao Shuang slammed the table and shot up from his seat. Big Su stood his ground. He kept talking.

'We all have problems. But I'm still coming to work every morning. The lake froze. Our boat went out of commission. We lost the family business. We ate congee and pickles for years until I found this job. I value my life. I would choose to live for ever, if I could.'

The man out of uniform had been silent this whole time.

'We have to be more sympathetic,' he said now and squeezed Big Su's shoulder. 'Rich people have problems too. Let's show the dead some respect.'

'She was born with a silver spoon in her mouth,' Big Su barked back. 'If you gave me her life, I'd be the happiest man on earth. I feel sorry for her family, I really do, but I have no respect for someone like her. In fact, the more I think about it, the angrier I feel. We try to save lives here and there are people out there just giving theirs away as if there's a donation box for it somewhere.'

'Listen to yourself,' the man replied. 'This is why you will never be rich. All you can think about with that big head of yours is how difficult your life is. You're a respected police officer with a stable job. Quit it with the whining.'

'I served this town all my life. Ever since I was a kid. First, the boat. Now, this.' Big Su pulled out his badge.

I couldn't believe what I was listening to. I wasn't even angry. It just felt absurd, watching these men argue over the death of a woman they'd never spoken to, seeing just how far their words landed from what really mattered. Miss Pan was never the point of this conversation. She simply gave these people an opportunity to let their miseries pour out of their mouths like piss. A thundering clarity reverberated inside my head.

Even with all her charm, in death Miss Pan had become nothing more than an excuse.

Gao Shuang pulled on my arm and we left. I followed him to an empty room two floors down. He didn't say anything on the way and flung the door shut behind him.

'It's all right,' I told him. 'They're idiots.'

He took a deep breath and sighed it out.

'I know,' he said. 'I work with them, remember?'

'Did you find out anything about Miss Pan while you were in the south?'

He fished out some photographs from his pocket. They were slightly wrinkled.

'Nothing suspicious in her belongings,' he said. 'These are some photographs of what she left in the room.'

He flipped through them and I recognised her make-up bag, the diary, the baby clothes, the camera – nothing I hadn't seen already.

'Did she write anything in the diary?' I asked.

Gao Shuang shook his head. 'And the film in the camera hasn't been used.'

'Can I keep these?'

He organised them into a pile and handed them to me. I looked at her gloves on my hands. Less than a day after her death, Miss Pan's existence had already become defined by what she'd owned.

'Do you also think that she was a murderer?' I asked. 'To be honest, I don't know what to feel. I don't know whether I should be sad or angry. She seemed so normal last night.'

Gao Shuang sighed and patted me on the shoulder.

'Do you think it was a moment of confusion?' I asked him. 'Maybe if I'd stayed, I could've stopped her. The baby could've been born, at least. Maybe that would've convinced her to live.'

'You can't let yourself think like that.' He perched against somebody's desk. 'A few years ago,' he said, 'I was assigned to a case. A sixty-year-old man tried to kill himself by inhaling gas. His brother found him unconscious and called the police. I spent a lot of time at the hospital with him. I didn't have to go as often as I did, but we developed a bond, and I liked listening to him talk.'

'I remember that. You said he was a professor at the University.'

'Anthropology. The more we talked, the less I was able to understand why he had tried to die. He seemed like such a reasonable person. So I asked him.'

'You asked him why he wanted to die?'

'He told me that he first began studying anthropology because he was young and angry and fed up with the world. He thought that we are an unstable society, like a pot of water that is ready to boil over. He wanted to know how we arrived here and he wanted to understand why he was so angry about the way things were, where this anger came from.'

The door opened and a man dressed in uniform walked in. He hurriedly leafed through some files in the cabinet by his desk, pulled a folder out and left again. Even in his rush, he remembered to close the door behind him.

'And then?' I asked.

'He tried all sorts of things to defuse this anger,' Gao Shuang continued. 'He'd got married to a beautiful woman, hoping that it'd help him learn to love and to care, but time went on and his wife was just that to him, a beautiful woman, nothing more. She'd managed to put up with him for eight years. Can you believe that? Eight years. Inevitably, she left him in the end, less beautiful than before.'

'So he tried to kill himself because he was angry?'

'The thing is, when I met him, he wasn't angry any more. In fact, he was as calm as the frozen lake. So I asked him, "Why try killing yourself at sixty years old? Why now? Why not when you were lost and confused?"'

'What did he say?'

'He said that after forty years of studying the human species, he finally understood that he was just another angry man among many other angry men. Ironically, that revelation was what ultimately brought him peace – the idea that he was just a part of this bigger, collective anger. A few months later, he attempted suicide again. And he succeeded.'

'I remember when that happened, you came over for a drink and we ended up having too much. Dong Ji and I were worried about how upset you seemed.'

'I was upset,' he said. 'I felt guilty for not being able to prevent it. But back to your question. I don't have an answer, but all this is to say that I don't think people take their own lives because they are confused. On the contrary, I think that they do it because they have too much clarity. It could be a moment of clarity in a bout of extreme confusion, if that makes any sense.'

He brushed some imaginary dust off his trousers.

'Maybe it's better to be a little confused then,' I said.

There was a purple clock sitting on the desk behind Gao Shuang. It read 3:50 p.m. I'd been there for over an hour already.

'I need to go back to the store soon,' I said. 'Something else happened last night but promise me you won't do anything rash.'

I recounted the events. I knew Gao Shuang would want to hear all the details, so I started from the hospital and combed my way through the course of the night. He listened with care. However, I stopped short at any mention of the light that came out of Driver Hua's mouth. I'd planned on telling him, yet when the chance was in front of me, I couldn't find the courage or the words. My hesitation was perplexing. After all, I'd been able to tell Dong Ji so easily.

'Do you have his licence plate number?' he asked after I finished.

'I can find out.'

'Don't worry about it. Just tell me his full name and I'll track him down.'

He seemed uncharacteristically calm. Perhaps he was doing it for my sake. He had matured, I realised. When Dong Ji and I were younger, Gao Shuang would get into trouble all the time. Whenever somebody wronged us, he'd go and beat them up, which often ended with him getting lectured by his superiors. He was smart about it. He'd always avoided getting into fights while he was in uniform, so the punishments were never too severe.

After our talk, Gao Shuang insisted on walking me back to the shop, even though he'd been gone from his desk long enough. The sky was dark and snowflakes were beginning to fall. After we arrived, I asked whether he wanted to come in.

'Dong Ji's inside,' I said.

Gao Shuang liked my sister. He always had; we all knew that. But he'd never confessed his feelings. The reason, I suspect, was a messy mixture of guilt and grief that he couldn't separate. In many ways, the day Ba died, Gao Shuang lost a father too.

Instead of responding to my invitation, Gao Shuang took a cautious step forward and gave me a hug. It was a soft, weightless embrace, so feather-light that our clothes absorbed most of it.

'I'm sorry,' he said. 'I'm sorry that it all piled up on you.'

The familiarity of his voice was like the comfort of an orange light at night. I lowered my head and tucked my face into his coat, feeling the icy fabric against my eyes. The webs of anxiety in my heart cleared a little.

When I entered the shop, the Su girl was standing across the counter from Dong Ji. At first, I didn't recognise her. I'd forgotten about her haircut. Her long braid had been chopped into a bob that sat right above her white scarf. I thought she looked better this way – lighter, more youthful. She reminded me of a moth.

'How did it go?' Dong Ji asked when she saw me.

She was searching for the price tag on a box of bird's nests. I hadn't ordered any since the last two boxes, so it must've been the one that Miss Pan had left with me.

'I saw your brother,' I said to the Su girl.

'He looks horrible, doesn't he?' she asked with an expression of worry. 'He hasn't slept in three days. I hope he can focus on work. I wish I could help him.'

'He seemed all right,' I said. 'He seemed like he could take care of himself.'

Before she could say anything else about Big Su, I turned my attention towards Dong Ji. I didn't trust myself to hide my distaste for her brother and what he'd said.

'Have you seen Gao Shuang lately?' I asked Dong Ji.

'I haven't had any time off,' Dong Ji said.

'You should have dinner with him.'

She raised her eyebrows as though to ask *What are you trying to say?*

The Su girl pulled out some money and started counting.

'It's a gift,' I told her. 'You don't have to pay.'

'But I brought money,' she said. 'It's expensive, isn't it? Why would you give it to me as a gift?'

'I meant that it is a gift I received from somebody else, so it doesn't feel right for me to charge you for it.'

Embarrassed by her own question, the Su girl looked down at the bills in her hands and blushed.

I explained the method of cooking the bird's nests in detail. She repeated every step back to me as I said them, taking care to show that she'd understood.

'Eating bird's nests is good for your skin,' Dong Ji told the Su girl. 'Makes it all peachy and soft.'

'That's nice,' the Su girl said. 'But I don't care too much about that. I just want to try it because I'm curious.'

'About what?' I asked.

'About what it's like to be a bird.'

'That's the strangest reason I've heard,' Dong Ji said.

'There used to be a large tree next to the dock,' the Su girl said. 'One morning, just when the weather was starting to get cold, a nest fell from it. And then, sometime later, another fell. Eventually, every day, a new nest would fall from the same tree. Sometimes, there would be a baby bird in it. I remember thinking that the mother bird who was building those nests really needed to learn how to do it properly.'

'Did you rescue the baby birds?' I asked.

'They were too young. All of them died. You can't keep a young chick alive. It needs its mother.'

For a short interlude, we all fell silent, mourning the baby birds.

'I cooked and ate one of the nests,' the Su girl said. 'I thought that was what rich people ate. It was all twigs and moss. I got so sick afterwards. But I really thought that was what people ate. I thought that tree was going to be my money tree. I thought the nests were like threads of gold falling from the sky.'

A soft sigh escaped her thin lips.

'Where do you think the birds used to fly to when they left Five Poems Lake?' the Su girl asked.

'They usually flew south,' I said.

'I wish I were a bird,' the Su girl said. 'Then I could fly anywhere. Have you ever tried to leave?'

I shook my head. Dong Ji didn't answer.

'I asked one of my co-workers the same question,' the Su girl said. 'She said she'd never want to leave. She said we live in a paradise.'

'Even with everything that's happened?' I asked.

'Yeah,' she said. 'Even with the sun fading and the birds leaving. She thinks that this place is a sanctuary amid the empty desert, a miracle. Though I'm pretty sure "paradise" was the word she used.'

'If that's the case,' Dong Ji said, 'then I can't imagine what hell is like.'

The Su girl didn't say anything. I'd noticed Dong Ji making such comments more regularly as of late. A bitterness imbued her voice; it was something I didn't remember being there just a few years ago. When we were younger, her desire to leave Five Poems Lake had been charged with an excitement. At times, when she felt particularly hopeful, there would even be a shimmer of joy that she couldn't conceal entirely. Now, all of that was gone, replaced by a shadow, like a cluster of clouds moving in towards her.

'I think my brother needs something stronger,' the Su girl said, changing the subject. 'He sleeps so little now. He's beginning to see things.'

'See things?'

'He saw cats hiding in the room last night. The night before, he saw black umbrellas hanging from our ceiling. And the night before that, he saw the boat . . .'

She sighed.

'It's bad,' she concluded.

'You should take him to a doctor,' I said as I grabbed a roll of string from the bottom drawer of the wooden cabinet.

The Su girl had wanted to say something else, but she must've sensed that I was not keen on chatting more about her brother. Dong Ji measured and cut a length from the roll. She tied it around the box of bird's nests and made a handle out of it. The Su girl didn't use the handle and instead cradled the box in her arms like it was her baby. Watching Dong Ji manage the shop, it was as though she'd never moved away.

After the Su girl left, I called Driver Hua's boss to ask whether he'd shown up to work. Before I even said anything, he began firing questions at me, most of them variations of the same thing I'd wanted to ask him. *That stupid son of a bitch*, he called him. According to him, nobody was picking up the phone at Driver Hua's home. His wife had moved back to the countryside a while ago. What frustrated my supplier the most was the thought that he had to look for another driver at such short notice, not to mention another truck. Apparently, Gao Shuang had already called him enquiring about Driver Hua's whereabouts, which must've displeased my supplier even further. He was a busy man, he told me, and didn't have time for this. Then, upon realising that I had no useful information to offer, he hurriedly said he was sorry that my next delivery would likely be delayed and ended the call.

A week later, we finally had a sunny day, in the sense that there wasn't a single cloud in the sky. The lingering symptoms of my cold had gone away entirely. I was on my way back from the market when I saw Gao Shuang walking out of the police station. I had a small bunch of spring onions sticking out of my tote, which had given me an unfounded feeling of pride, as though despite the cold and drab weather, the labour of running a shop all alone, the looming disappearance of the sun, the incident with Driver Hua, the death of Miss Pan, the this and the that – despite all those things – I still possessed the strength of will to be kind to myself and add spring onions to my food. I was aware that the rationale was rather nonsensical, but I was feeling quite sure of myself when I spotted Gao Shuang waving his big hand at me from the metal gates.

'Had lunch yet?' he asked as I approached, his face beaming. When he smiled, his right eye would always become smaller than his left.

'I bought some skewers,' I said. 'They killed a sheep today and I got there early. They ran out within ten minutes.'

'Come and eat with me in the station.' He turned and started walking back.

'It's too dull in there,' I told him. 'I'm going to eat in front of the shop. It's such a nice day today.'

'Fine. I'll come with you then.'

'You know Dong Ji isn't there, right?'

'So?'

'Just letting you know, in case you have something better to do.'

'I don't. Nothing's more important than lunch.'

He followed me a few steps to the shop.

'Guess what?' he said, grinning.

'What?'

'You'll never guess.'

'Then tell me.'

He reached into his pocket, fished out a car key and dangled it in front of my face.

'They finally gave you a car, huh?'

'A seven-seater!' he said proudly.

We went inside the shop and pulled some stools out onto the kerb. We sat down, hunched over, with our lunches in our laps.

'I thought the seven-seaters they have are all really old,' I said.

'All the cars we have are old. Plus, it is big. I like big cars.'

'Why? So that you can go around and intimidate civilians?'

'So that I can protect you.' He nudged me with his elbow.

'I don't see how that has anything to do with having a big car.'

'You wouldn't understand,' he said. 'Forget it. Let's get some sun while we can. Breathe in the fresh air.'

He straightened his back, turned his face towards the sky and inhaled deeply. His chest expanded like a balloon.

'What are you?' I asked. 'A sunflower?'

He blew the air out of his lungs.

'I'm far more useful than a pretty little flower,' he said.

'Fine, fine, officer. What's for lunch?'

I pointed at his metal lunchbox with my chin. He took the lid off and held the box out to show me. Some fried noodles sat limp and wet.

'Do you want me to go inside and heat that up for you?' I stood up. 'It won't take long.'

'It won't help much,' he said, tugging on my arm to sit me down again. 'Eat your precious lamb before it goes cold.'

I took a skewer out of my paper bag and pulled a piece of meat off with my teeth. Some of the sauce smeared onto my cheeks. I gave Gao Shuang a skewer. Between bites, we chewed with our faces turned up towards the sky, letting the light wash over us. It was warm on the skin. If I used my imagination, I could even convince myself that the sun was still whole. I unwrapped my scarf and took a deep breath, tasting the lamb in my mouth, the spices that stuck to my gums.

'I found the truck,' Gao Shuang said, putting an end to the pleasant moment. 'It was burned down. But no sign of Driver Hua.'

My first thought was that it must've been Driver Hua's little sun that had set fire to the truck. But was that even possible? Did it have the capacity to burn the things it touched?

Frankly, I was hoping that Gao Shuang wouldn't

bring up the subject of Driver Hua. I'd spent the past week finding every excuse to distract myself. With each night that'd passed, my memories were losing their sharpness, giving me hope that maybe I'd forget about all of it sooner than I'd expected. But sometimes, out of nowhere, a detail – the garbage bag in the corner, the chipped brick on the wall, the shape of the moon, the smell of wet tobacco – would hurl itself upon me and the entire night would come back with a biting lucidity. It was as though I'd been marked by something filthy, like dog urine, and no matter how I cleaned myself, the scent was always going to find ways to come through.

'When the sun is gone,' Gao Shuang said, 'let's move somewhere else. With Dong Ji as well.'

I looked over. His eyes were still closed. He was tall and well built, with a large face and pudgy cheeks, and his features were clean and delicate. It wasn't a handsome face, but I found it pleasant to look at.

'If the sun is gone here,' I said, 'then it'll be gone everywhere. Moving away won't help.'

'I know that.'

'Also, we don't know what happened to all those people who left Five Poems Lake. They're probably all dead in the desert somewhere.'

'I know that too.'

'What's the point of moving away then?' I held out a skewer. 'You want another one?'

He shook his head.

'You're all skin and bones,' he said. 'You're not going to live a long life if you don't eat well. You need to learn from Dong Ji and eat more. She could eat a hundred dumplings if you let her.'

'If the sun disappears, we'll all die. It won't matter how much we eat.'

He opened his eyes and turned towards me.

'But I will survive longer than you because I have more fat on me,' he said.

'Are you not afraid, Gao Shuang?' I asked. 'I'm afraid.'

'Of death?'

'No.'

'What is it then?'

'I don't know. I don't always know what I'm afraid of.'

He gave me a bemused look.

'Sometimes,' he said, 'I have no idea what's in that head of yours.'

With his chopsticks, he picked up some noodles that were all clumped together and took a bite.

'I'm afraid of being powerless, I guess,' I said.

'What's that got to do with the sun?'

'I don't know,' I said. 'The sun is what gives us life, but it can also take it away. I guess I just don't like that. It doesn't sit well with me. The simplicity of it, I mean. It makes everything we work for seem small and pointless.'

Gao Shuang spared my words a moment of contemplation before stuffing the last mass of noodles into his mouth.

'I actually quite like that thought,' he said as he chewed.

'You do?'

'The sun doesn't have a will. It's there. It's neutral. It doesn't *want* to give us life nor does it *want* to take it away. It just *is*. We have to accept it, live with it, and die with it.'

'Who replaced Gao Shuang the police officer with Gao Shuang the philosopher?'

A smile swiped across his face.

'I've always had this attitude,' he said. 'You've got to take the world as it is. There's nothing wrong or shameful in that. Your father used to tell me off when I said these things. He would say, "The job of a police officer is to accept justice and only justice, even when you don't have the power to uphold it."'

'Do you still feel the same way?' I asked. 'Do you think he was wrong?'

'I don't think he was wrong,' he said, 'but I'm not sure he was entirely right either.'

I looked over at the pot of camellia sitting just behind the glass door. Even though the flowers had bloomed later than usual this year, all of them had fallen by now. Most of the plants and trees around Five Poems Lake had turned into the dark muddy colour of kelp – some were even black – but very occasionally, you'd come across one like this camellia plant that had retained its bright green colour. It was really quite astonishing that

Dong Ji and I had managed to keep it alive after Yeye died. Neither of us were very skilled gardeners. To be honest, before taking over the camellia, I'd managed to kill every plant I'd owned. I'm not one to be superstitious, but there were times I really wondered whether it was Ba's will that kept the flowers blooming every year without fail.

Ba had plenty of faults, but everyone agreed that he was a good police officer. Even when I was a child, I could see that. It was like he'd been born for it. He had a sharp eye and a trustworthiness, and if anyone in the world really knew what justice meant, it must've been him. But when it came to looking at himself, it was as though he was blind. He couldn't figure himself out at all. As a matter of fact, he didn't even try. I found it impossible to understand how he could be so in tune with everything but completely oblivious to himself. I was the opposite – so aware of my own existence that sometimes I couldn't make sense of the world at all.

'You know what else your father used to tell us?' Gao Shuang said.

He must've noticed that I was in a daze. His voice was loud.

'What?'

He closed his lunchbox and stood up.

'What?' I asked again.

He laughed.

'Tell me!' I smacked his arm.

'OK, OK, I'll tell you. Didn't they teach you not to hit a police officer?'

'Who was supposed to teach me that?'

He straightened his face and pointed at my lunch.

'He always told us that this skewer stand is filthier than pigpens. That guy who does the grilling doesn't wash anything. Not even his own hands after he pisses.'

He laughed some more and ran off before I could hit him again.

I watched him as he reached the metal gates of the station. He gave a dramatic wave before disappearing behind the security guard booth, and in that moment, his face was like the sky.

3

I awoke thinking that it was dawn, but turned out it was already eight in the morning. If Gao Shuang hadn't been ringing the doorbell, I would have still been asleep.

'It's so dark,' I said as I opened the door.

I must've looked rather unseemly. I hadn't even washed my face.

'Thank God. You had Dong Ji worried. She called me to come check on you. What took you so long? Did you see the weather forecast?'

'Why? Did another piece of the sun disappear?'

'Twenty per cent!' he said. 'In fact, they said it was slightly more than twenty per cent that went away overnight. But there's something else. I don't even know where to begin. You better see for yourself. I just came here to make sure you're safe.'

'Twenty per cent? Are you sure?'

'Take a look at the sky.'

I charged outside, still wearing my pyjamas. The cold was like nothing I'd felt before. The ground was obscured under a thick layer of snow. Was this what it was like to be living at the end of the world? I looked towards the east and saw the pathetic excuse of a sun. It was no longer a recognisable shape, more like a small red smudge above the police station, as though with the rub of a finger, it would disappear.

I went back inside and turned on the television. Gao Shuang followed me. They seemed to have assembled a hasty panel of people, all men. The weatherman with the ties wasn't on. A woman nicknamed Hong Dou – or Red Bean – was moderating the discussion.

Twelve years ago, when the sliver of the sun first disappeared, Red Bean had achieved some sort of superstar status among the people of Five Poems Lake. All of us would turn on the television first thing in the morning to listen to what she had to say about the sun. As the sun continued to fade, however, many people stopped watching television. They didn't want to hear about the sun every day – or so they'd said in a survey conducted the year that Red Bean's popularity waned substantially.

'What kind of cold weather are we talking about here?' Red Bean asked. 'Just how cold will it become? It's only been five days since the eight per cent fading.'

'It's difficult to say,' a square-faced man said. 'Nothing like this has happened before.'

'We'll just have to wait and see,' another man said feebly. He must've been over eighty years old. 'We have to be resilient but not rigid. Or else we will break.'

Red Bean fired back, 'But being resilient is not enough any more, is it? What about food? Will our farms be able to survive this?'

'We must try our best to be prepared,' the square-faced man responded, his hands making too many gestures as he spoke. 'We need to make sure we can get supplies to every household. Heaters. Blankets. Extra clothes for those who work outdoors. Everybody has to stay warm.'

'Why has the vanishing sped up so much in these past days?' Red Bean asked. 'How much time do we have left before the sun completely goes away? Looking at the current trajectory, could it be weeks? Days, even? What happens afterwards?'

'There is no pattern so far,' the old man said. 'We can't be sure. But we shouldn't say it's the end.'

'I'm not saying it's the end,' Red Bean said.

'Yes,' the square-faced man said. 'Saying it's the end would be premature. Not to mention too pessimistic.'

'What about this?' Red Bean said. 'Here's a rather disturbing scene captured last evening.'

'Watch this,' Gao Shuang said.

A video clip began playing. In the darkness, there were three people standing by the lake. To my shock, they were just like Driver Hua, with fiery lights shining from the place where their heads should've been. From their clothes, it seemed that they were two women and one man. They stood in place, turning their bodies, as though they were looking for something, or somewhere to go. I would even describe the way they carried themselves as calm. Someone could be heard screaming off camera. The video ended.

'What is this about?' Red Bean asked.

'We can't rule out the possibility that it is somebody's idea of a joke,' the square-faced man said.

'This is troubling,' the old man said. 'Even if it is a joke, everybody I've seen this morning has been talking about it. We can't let unfounded rumours spread. The police are looking into it. They are the experts.'

'It's not a joke,' I said. 'The night Miss Pan died, Driver Hua turned into one of those too. A light came out of his mouth and wrapped around his head, just like that. I saw all of it.'

There was a beat before Gao Shuang reacted. I couldn't bring myself to look him in the eye, after having withheld this information the last two times I'd seen him. It had not been my intention to lose his trust, but even so, I had no excuse.

'I know,' he said. 'Dong Ji told me on the phone. She said that you saw it happen.'

He said this without any anger in his voice. Relieved, I looked at him and nodded.

'There are more,' he said. 'I shouldn't be telling you this, but to hell with it. We received a few calls throughout the night reporting similar incidents. From what I know, the other districts have got multiple reports too. They're calling them "Beacons".'

I opened the curtains and peered out the window. The back alley was dim and silent. Only one pair of footprints could be seen in the snow. *Beacons*, I thought. *How poetic.*

'Do they stay alight?' I asked. 'Or do they burn themselves up like matchsticks?'

He blinked at me, as though those were not the questions he'd expected.

'If they do burn themselves up,' he said, 'it hasn't happened yet. Not according to the reports anyway.'

'Do you think Driver Hua might've been the one who burned the truck down?'

'Is that what you think?'

'He told me once that he didn't want to just be known as a truck driver,' I said. 'Maybe that's why he burned it down.'

'So you think these Beacons . . .' He paused. 'You think they still have thoughts like that?'

I turned again towards the television. They were showing the video again, pausing every few seconds to zoom in on the Beacons.

'I don't know anything,' I said.

'You know more than all of us,' Gao Shuang said with a sudden sureness in his voice. 'Even with all the calls we got, it seems that no one has witnessed the exact moment it happened. You're the only one who can tell us how it happened and testify that it's real.'

'It *is* real,' I said. 'Didn't you see the video?'

'What I'm saying is that you should tell the police officially.'

The suggestion floated out of his mouth like air, almost like it didn't hold the immense weight that it did. I told him I'd have to think it through. I had no problem reporting Driver Hua for all that he'd done, but I felt sorry for his wife, and this unexpected sympathy held me back. How would she feel, knowing that she was married to a man who'd lost himself so irreversibly that even his head was gone?

After Gao Shuang left, I opened the shop. No customers came that morning. It must've been because of the Beacons – everyone was afraid. In the afternoon, since I still hadn't seen a customer, I locked up and took a bus to the Western District. I decided I'd discuss with Dong Ji about Gao Shuang's suggestion to file the report. She always knew what to do.

More people were out and about and working in the Western District. It didn't really surprise me. Maybe that was why those who were in this area were a good deal wealthier.

A woman who seemed to have had too much to drink caught my attention while I was making my way to the wellness parlour from the bus stop. She was stumbling towards me and I recognised her as Red Bean. She wore a black coat and her face was still made-up for TV, giving off the impression that it had nothing to do with her body.

She was trying her best to avoid bumping into people, though I couldn't say that she was very successful. She held a bottle of liquor in her hand, swinging it like Driver Hua did when he'd walked into my shop. I stepped out of her way as she dragged herself past me, rocking from side to side, and collided with the man who had been walking behind me. The man recognised her and called out her name. She smiled with her made-up face and waved at him. I watched her lean against the corner of a building for a second and then disappear around it.

Dong Ji met me by the back door of the parlour, where the deliveries were dropped off and the garbage was picked up. She was wearing a milky white shower cap over her hair and held a blue duckbill clip in her hand. Snowflakes melted onto her cap as they fell.

'Can you believe this?' Dong Ji said. 'People are still coming in for massages and facials. They must've all lost their minds.'

'I guess we're not ready yet,' I said.

'You don't say.'

'Maybe I should tell the police everything,' I said. 'Gao Shuang told me that they could use the information I have.'

'Do you want reporters showing up at the shop? Or worse, bad people who might hurt you?'

'I thought you were adamant that Driver Hua should answer for what he did.'

'I was. I am. But these Beacons . . . Honestly, I didn't believe a word you said when you told me. Whatever happens, I just want you to stay out of trouble.'

She pulled a neatly rolled cigarette out of a pouch and clasped the duckbill clip onto the end. She smoked without touching the cigarette with her fingers. She'd often complained that the scent of tobacco was difficult to wash off.

'I want to talk about something with you,' she said.

'Something else?'

She brought the cigarette to her lips, took a drag and held the smoke in her lungs. When she breathed out, she turned her head so that the smoke travelled in the same direction as the wind.

'I want to dig up the urn,' she said. 'Look at Ba's ashes.'

'What? Where does this come from?'

'We never opened the urn, right? I want to see for myself.'

'But what about the camellia?'

'It's what's under that matters, isn't it?'

'But isn't it disrespectful?' I asked.

'To Ba? Or the plant?'

'I don't know. To both.'

'You think so?'

'I don't know.'

'Yeye would never let us,' she said. 'If he were here.'

'That I do know.'

A truck came and began to back into the parking

space. It made an irritatingly loud beeping noise as it reversed into position.

'It's freezing,' said Dong Ji as she hopped up and down a few times.

I had become rather attached to the camellia plant after all these years of caring for it. It was forked at the trunk and one branch was slightly smaller than the other. I'd always thought that the two branches felt symbolic of Dong Ji and me.

'I quite like that camellia,' I said.

'I like it too. It reminds me of the grassy area that used to be around the basketball court. There were lots of camellias there.'

When she was in high school, she'd often drag me along with her to watch the boys play basketball. It was an open secret that she had a crush on one of them. The court was by the lake, and next to it was a small patch of grass. We'd sit on the grass and eat our snacks. The area had all turned to dust by now.

'But the plant is going to die,' Dong Ji said. 'I don't want Ba to be stuck under a dead plant, with nobody to care for him. The sun could disappear any day. We could be gone too.'

We both looked up at the sky. The sun was so weak that it was possible to look directly at it without hurting our eyes. Dong Ji unclasped the clip and the burning cigarette butt fell into the sludge on the pavement.

'What do you want to do with the ashes?' I asked.

'Nobody swims in the lake any more,' she responded immediately. She must've already run this scenario through her mind. 'We can break a hole in the ice and pour the ashes in.'

I'd never given much thought to what we'd do with Ba's ashes if the sun were to disappear completely. It is a common belief that after death, our souls descend deep into the water, beyond the bed of the lake. If that is true, wouldn't we all be reunited there anyway? If, on the other hand, there is nothing after death, then why did it matter?

Dong Ji didn't think that way. She wasn't so naive that she'd cling to the illusion that there would be significance in the things we left behind, but in the little ways she could, she wanted to put care into life and death – her own and that of others. If I were a log that washed along with the currents, then she would be a fish that swam in the stream. Neither of us could escape the water, but I was the one who didn't fight my fate.

'All right,' I said. 'Let's do it.'

'Aren't you going to ask me why I'm deciding to do this now?'

'Are you going to tell me?'

'I guess I just want to see,' she said. 'With my own eyes. I want to be sure. I've been thinking about it for some days now, ever since you told me about Driver Hua. I didn't believe you, but I couldn't get it out of my head. And then, seeing the Beacons this morning . . . I don't know. It just all made me think of Ba, somehow.'

'The Beacons? What have they got to do with Ba?'

'Don't take this the wrong way, and I do despise Driver Hua for what he did, I really do. But I'm just starting to think that maybe the appearance of the Beacons isn't such a terrible thing to happen to us.'

'What do you mean?'

'Because at least *something* is happening. Disrupting the trajectory of things. So I'm thinking that maybe it's not a bad thing.'

'You didn't see Driver Hua, Dong Ji, there's nothing good about what happened to him.'

She held her breath for a moment before letting the air out with a loud sigh.

'You're right,' she said. 'There's nothing good in this godforsaken place.'

She gave me a hug. Her shower cap scratched against my ear. The scent of all her creams and oils made her smell like a freshly bathed baby.

'We'll do it tonight,' she said. 'After I get off.'

'Tonight?'

'Are you scared?'

I shook my head. 'It's Ba,' I said.

'Look at you,' she said. 'So brave. To be honest, I'm a little scared.'

She patted me gently on the back and told me she'd see me again in a few hours. I watched her take off her shower cap and sniff the ends of her hair to make sure they did not smell like smoke before tying them up into a tight bun. When I left, the truck was still parked. The driver was taking big bites out of a spring onion pancake. It was a woman. She had on a bright yellow beanie and her hands that were holding the pancake were red from the cold.

At the bus stop, I saw Red Bean again. She was sitting on the ground and leaning against the bench. Her eyes were staring forward at the lake. It was two in the afternoon and most people were finished with their lunch breaks, so there weren't any others waiting. I walked up.

'Do you need any help?' I asked.

She saw me and pushed herself up from the ground, wobbling on her feet.

'I've got a terrible headache,' she said. 'Can you buy me a bottle of water? I have cash.'

She tried to unzip her coat pocket, but the zipper got caught on a piece of fabric.

'Don't worry about it,' I told her. 'Wait here.'

I found a convenience store nearby and paid for two bottles of water. When I returned, she was lying on the bench with her legs dangling off the end. I handed her a bottle and watched her sit up and drink half of it in one go. Her mascara had smudged. Close up, I could see her enlarged pores through the thick layer of foundation. The red lipstick made her look like she'd eaten an entire raw pig liver. We sat together and sipped our waters until the bus came. I got on. She said she wanted to stay behind.

Dong Ji didn't come over until almost midnight. Most of her customers had to work during the day, and after they got off, they had to delegate time for eating, so the parlour was always the busiest in the evenings.

Before she arrived, I'd already moved the camellia plant from the front of the shop to the living room. It was heavy but not so much that I couldn't push and drag it along the floor by myself. Dong Ji squatted down next to the plant and pressed her fingers into the surface of the soil. The damp soil stuck to her skin.

'We should get started,' she said.

I grabbed a trowel and handed it to her.

I held the plant while Dong Ji took care of the digging. She remained vigilant, making sure to keep the roots intact. We decided that we'd replant the camellia in the pot after we removed the urn. After a fair bit of digging, the plant began to sway.

'Pull it up,' Dong Ji said. 'Slowly.'

Gripping the forked branches, I tried to lift the plant up. Leaves fell off like pieces of paper.

'Stop,' she said. 'Stop. Hold on. There is more in the soil.'

I had to keep the plant levitated while she continued to dig.

'It's heavy,' I said. 'Hurry up.'

My lower back was beginning to hurt from the

weight. The branches scratched against my arms and face, so I had to lean my head back, which ended up putting more strain on my spine.

'OK,' she finally said. 'Lift!'

Dong Ji threw the trowel to the side and helped me raise and move the plant from the pot to the back door. We carefully laid it down on the floor.

'You all right?' Dong Ji asked.

'Yeah.'

'Your face got scratched up,' she said as she reached out to touch my forehead. 'It's a bit red.'

I leaned backwards to avoid her hand.

'I'm fine,' I said. 'Let's keep going.'

We dug some more in the pot and, after a few minutes, found the plastic bag. Somehow, even though it couldn't have gone anywhere, we both froze for a moment when we saw that the urn was still there, at the bottom of the pot.

'Sorry to disturb you, Ba,' Dong Ji said.

She pulled hard on the plastic bag and unearthed the porcelain urn.

Even after all these years, the permanence of death still eluded my understanding. The urn, stained with the distinct yellow of old glaze, made me think about how being dead was not the same as being gone from this world. Pigments remain, some brighter than others. Ba would never live again, yet he wouldn't completely disappear either, like those traces of him inside the police station. It sounds rather simple – there is nothing philosophical or complex about it – but when a human life was involved, the simplest facts always seemed more complicated.

Dong Ji was the first to move. She took the urn out from the plastic bag, releasing me from my thoughts. We cleaned the surface with a wet towel. The yellow glaze bled into the glossy white porcelain like tea stains.

'You do it,' she told me. 'Are you ready?'

'OK.' I took a deep breath and wiped my hands on the towel. 'OK.'

We counted down from three and then, eyes closed like a child, I popped open the lid.

When I opened my eyes again, I saw that inside the urn was a pine-green silk pouch with golden flowers woven into the fabric. It was an exquisitely crafted thing, nothing like I'd ever seen before. The threads of gold were as thin as hair and shimmered in the light as though they were wet.

'What's that?' I asked. 'What about the ashes?'

'Must be inside.'

Dong Ji took the pouch and looked at me before untying the knot that was sealing it. In it, we saw Ba's ashes, a light grey, with some specks of black that rose to the surface as Dong Ji squeezed the outside of the pouch and moved the fabric around. The ashes had no scent, which made me feel uneasy and even a little confused. When he was alive, Ba had always smelt like smoke. I must've expected his ashes to carry the same scent; to find that they were so devoid of anything that resembled Ba was upsetting. It made me feel as though whatever part of him that I'd thought had remained behind was, in fact, not there at all.

Dong Ji held the pouch out in front of me and, as I leaned over to take a better look, I saw the folded corner of a piece of paper sticking out from the ashes. I touched it and, judging from its thickness and glossy surface, I realised that it was a small photograph. I clasped the corner and pulled on it. With some resistance, the photograph slid out, revealing an image of what could only be described as a Beacon.

The photograph was taken from a low angle, as though the person behind the camera was sitting on the ground. The flash was bright and the background was entirely black – it must've been taken at night. In the centre of the photograph was the white, skinny figure of a man with a bright light shining atop his shoulders.

Upon seeing the photograph, the pouch fell from Dong Ji's hands and some of the ashes poured out onto the floor.

'I'm sorry,' she said. 'I'm sorry. I didn't mean to drop it.'

She stood up and turned right, then left, and then right again, like she didn't know what to do with her body.

'Fuck,' she said. 'Fuck. Where's the broom?'

'The broom's dirty,' I said. 'I don't want to sweep Ba's ashes up with a dirty broom.'

'You're right, you're right. What was I thinking?'

I began sweeping the ashes into a pile with my hands. She grabbed her purse and took out two cigarettes. Her hand was still unsteady when she passed one over to me. We were silent as we smoked and gathered the ashes and scooped all of it back into the pouch.

'Be careful not to get any cigarette ashes mixed in,' I said. 'Though Ba probably wouldn't mind some tobacco after all this time.'

'The photograph was taken with Ba's camera,' Dong Ji said, ignoring my attempt at humour.

'How do you know?' I asked.

'Because of that,' she said, pointing at a thin white line running down the side of the photograph. 'The light leak.'

I could feel my heart expanding in my chest. What on earth was going on? Were there Beacons when Ba was alive? Did Yeye put the photograph inside the pouch and if so, what else did he keep from us? Was that why Yeye handled everything by himself?

My mind couldn't follow through with any of these questions. I'd try to focus on one, but soon another would take over. It was as though there were a thousand people talking to me at once. Dong Ji sat across from me and switched between smoking and holding her breath.

'This is our chance to find out,' she said finally. 'About what really happened to Ba.'

'About how he died?'

'Maybe it had something to do with this man in the photograph.'

'What are you saying? You think this man killed him? But they said he drowned in the lake. They found his body—'

'How could you say those words so easily? He *died*, his *body*, things like that. It's like you're talking about a stranger.'

With that, we retreated back into silence.

After some time, I said, 'There are just too many things I don't know. Things I keep thinking about but can't seem to understand. Questions that keep bringing up more questions. The sun, the Beacons, why Miss Pan jumped off the rooftop. And now this.'

'It's always been an unknown,' she said.

I didn't say anything, because she was right. The secrets surrounding Ba's death were like a giant canopy that had loomed over us for so long that I'd forgotten about the clear skies above.

'Why did you want to look at Ba's ashes?' I asked.

'I told you,' she said. 'Because we'd never seen them before.'

'You said it had to do with the Beacons. It can't all be a coincidence. The Beacons, this photograph, your suggestion to dig up the urn. Why?'

She avoided my question.

'If we want to figure anything out,' I pressed, 'you'll have to tell me whatever it is you're keeping from me.'

She didn't speak for a moment. It wasn't hesitation that I sensed from her, but rather a strong resolve that gave a tangible weight to the silence.

'Ba came home,' she said. 'It was the night before he died. You were asleep, and it was just me and Ba. I'd been worried about the sun disappearing. I couldn't sleep. And Ba, well, he mentioned that we could "make our own suns".'

'Make our own suns?'

'If it'd come out of anyone else's mouth,' Dong Ji continued, 'I would've forgotten about it. But Ba never talked that way. He didn't like to speak in such vague language. He never made up those little stories parents use to comfort their children, to explain things that would've otherwise been too difficult for their children to understand. He was always honest with us. Too honest, most of the time.'

She tried to smile.

'I just remember thinking that it was odd,' she added.

'Maybe he felt like joking that night.'

'Maybe. But why the night before he died? I just have a strange feeling about all of this. Look at this photograph. Don't you think that he might have been talking about the Beacons?'

'There are so many things that he could've meant. He could've been talking about, I don't know, finding hope from within yourself, or maybe he was just telling you to draw a sun with crayon or mould one out of clay. Make our own suns, you know? It could've been so many things, Dong Ji.'

While I was talking, she just kept shaking her head.

'No,' she said. 'It's not any of that. I know it. You weren't there, little sister.'

She pointed to the photograph.

'How else do you explain this?' she said.

I thought for a moment. 'Why didn't Ba tell the police about this Beacon? It seems so unlike him, to keep it a secret.'

'That's why we have to find out,' Dong Ji said as she stared down at the photograph.

Her expression made it difficult to decipher whether she was deep in thought or entirely spaced out.

'Who do you think this is?' she asked eventually.

'Doesn't look like anybody I know,' I said, shrugging.

Dong Ji let out a chuckle as her eyes glided up from the photograph and met mine. They reassured me that she, too, was nervous about all that was happening around us.

'After that,' she said, 'we should get out of here. We just have to decide on it and go through with it. Before it's all too late.'

I'd never liked to hear her speak about her own plans as though they were shared between us. When I was young, her doing so had always made me feel like a child, and that is rarely something one wants to feel, not even when it is true. Later on, whenever she would make suggestions like 'Let's get out of here' or 'We should seriously think about leaving', I'd sense a selfishness embedded in her words that she wasn't aware of. In fact, she viewed her plans as the exact opposite, as if including me meant that she was able to construe her own desires as something more altruistic, prove that she was not making such decisions solely for herself. But by always putting me first, she had unknowingly backed me into a corner where living my own life would've been a betrayal of all her sacrifices. It came from her subconscious – from a good heart, no doubt – yet it angered me to no end. The burden of love was a heavy one at times.

I picked up the porcelain urn and tried to see a reflection in the dull surface. I felt compelled to stop her, to talk her out of this plan. Leaving had always been her dream, but dreams were supposed to make one happy, and I couldn't imagine anything but misery waiting out there in the endless desert.

While lost in my own thoughts, I'd gathered the loose soil around me into a pile. I turned my palms face up. The soil that stuck on them reminded me of the events that night in the back alley. Driver Hua, the moon, his bare thighs, soggy tobacco. I brushed my hands together.

'I think about Ba more than I think about Yeye,' Dong Ji said, 'even though it was Yeye who raised us. I used to think it was because Ba was gone so suddenly, but now I'm starting to think that it's just because we never saw it happen with our own eyes. I reckon there are some things that you just have to see, and this happens to be one of them.'

'Even eyes aren't entirely dependable,' I said.

'Speak for yourself. My eyesight is sharp and precise.'

She flicked my forehead and gave a weak laugh when I made a face.

'After my mother left,' she said, her expression serious again, 'I tried to believe that she was dead, but I just couldn't convince myself. So I began thinking that maybe she was alive, but I couldn't confirm that either.'

Her words slowed down.

'They all took their responsibilities, crunched them up into a heavy ball, and just threw the ball back at us. Ba, Yeye, my mother.'

She took another cigarette out and looked down at the silk pouch in between us. The lamplight smeared over her back, casting her black shirt with a muddy colour.

'Ba's gone,' she said, flattening out the photograph. 'I know nothing will change that fact. I'm not so naive. But we have to know the choices he made. Did he choose to leave us? That's what makes a difference.'

A sadness rose to her face.

'I do wish that we could've said goodbye,' she said, rolling the cigarette between her fingers. 'To Ba, to Yeye.'

'I talk to Ba sometimes,' I said. 'When I'm cleaning the altar and changing the fruit.'

She lifted her eyebrows in surprise.

'Never took you as somebody who would do something like that,' she said as she gave my shoulder a soft punch.

'It just gets so quiet here. Especially when I'm cleaning the altar.'

'We never get to say goodbye,' she said. 'Even when we know they are going to die, we still don't really have the wisdom to accept that any conversation could be the last. So we wait until they're gone and then it's too late to say anything.'

Was it really wisdom that we didn't have? I thought. *Was it not courage?*

'Sometimes,' I said, 'goodbye isn't something you say out loud.'

'You know,' she said as she turned towards the altar, 'maybe the thing that makes us family is that we can never really say "this is where we part", not even in death.'

Outside, somewhere in the distance, a dog was barking. It sounded like the black mongrel that Old Li had taken in a few years ago.

'Should we tell Gao Shuang about this?' I asked.

'I want to avoid bringing up Ba's death,' she said. 'I don't want to open old wounds.'

'You won't be able to hide it from him for too long.'

'He's too good to be pulled into this,' Dong Ji said. 'It hurts me to think about the pain it would bring him.'

'I think I know what you mean.'

Maybe she was right in that it was better for Gao

Shuang to believe in the cold simplicity of death, but she herself didn't even believe in it. He did not need us to protect him from the truth.

'But I don't want to keep secrets from him,' I said.

'We will tell him once I think of a plan.'

She stood up, signalling an end to our conversation. We stored the pouch in the urn and placed it on the altar, taking away the incense burner and candles to make space. Then, we planted the camellia tree back into the terracotta pot. As for the photograph, after some discussion, we decided that we should put it on the altar as well, behind the placard. The golden characters 'Dong Yiyao' on the placard brought back memories of Yeye and the days after Ba's death.

After Dong Ji went to bed, I stayed up to clean the living room. I enjoyed physical work. It would distract my body just enough that it couldn't react to the things going through my mind. This way, I could let myself think more freely, without the fear that my thoughts would paralyse me, launch me into a panic. The mind is a dangerous construct. What a paradoxical state of existence we're in, where our minds can stretch across dimensions, deep into reality and far into fiction, assign meaning to everything, yet our bodies are so small and limited that I can't touch the roof of my room. Even a marble could kill us if it hit the right spot.

FIVE POEMS LAKE, BEFORE DONG YIYAO'S DEATH

4

Dong Yiyao was sitting on a fire hydrant with his legs crossed and eyes gazing up at an apartment window on the third floor of a faded red building. There was a faint, white light glowing from inside. He'd gone through the rest of his tobacco already and did not know what to do with himself. He began biting his fingernails and chewed a corner off one of them. It smelt like mould. He deliberated making a quick trip to the store to pick up another pack of tobacco and some rolling paper, which should sustain him through the night, but if Ma Gang left the building while he was away, he would've spent all this time waiting for nothing.

For seven nights in a row, Dong Yiyao had been watching Ma Gang's apartment. On four of those nights, Ma Gang's lights had turned off around two or three in the morning. Dong Yiyao would stay for fifteen minutes or so afterwards to make sure Ma Gang was asleep before going home. On the other nights, the lights had remained on until dawn, so Dong Yiyao had waited until the sun rose. Then, to make sure that Ma Gang was not doing anything out of the ordinary, Dong Yiyao followed him all the way to work.

So far, Dong Yiyao hadn't been able to decipher a pattern. So, night after night, he waited and watched.

In the summer, evenings were the most oppressive. There was no wind – not even a breeze – and the temperatures at night remained just as hot as during the day. The sky was cloudless, making it hard to tell just how high up it stretched. The full moon was the colour of an oyster. It seemed as though losing that sliver of the sun hadn't made much of an impact on the weather at all. Dong Yiyao took off his shirt and used it to wipe the sweat off his forehead and neck. His undershirt was soaked through. He made a mental note to bring a towel the next day.

Dong Yiyao had begun his nightly watch shortly after Ma Gang reported that his wife was missing. *She must've been gone for days*, Ma Gang had told the police, and continued to say that it shamed him to admit that it'd taken him so long to notice. According to Ma Gang, it wasn't unusual for Tan Tan to spend her nights elsewhere, but he'd found it strange when she failed to show up for her weekly acupuncture appointment. Ever since Tan Tan's parents died from heart issues, she'd been particularly watchful of her health – obsessed, even – so it was unthinkable for her to have missed her appointment. She was terrified of dying.

Just a day into the investigation, after Dong Yiyao

had finished his preliminary inspection of Tan Tan and Ma Gang's apartment, Ma Gang came back to the station and decided that there was no need to look for his wife any more. He told Dong Yiyao that Tan Tan had left Five Poems Lake with her lover.

'I don't want to find her,' he insisted. 'Don't waste your time looking for her any more. Let them starve to death in the desert.'

When asked how he'd found that out, he began to cry. Between sobs, the only thing he said, over and over, was that he couldn't believe this had happened to him. He was so good to her, after all.

Ma Gang was a man who looked as though he'd never gone through puberty. He was short and so scrawny that Dong Yiyao wondered whether he could even pick up a shovel. Because of his childish appearance, he quickly became the target of jokes among the more unkind officers at the station.

'No wonder his wife was cheating,' one of them said. 'His thing probably doesn't work.'

'Does he even have a thing?'

It took a few days and a fair amount of lecturing, but Dong Yiyao eventually managed to get them to quit, and the case was closed before the investigations even began. For the police, at least, it was one thing off their plate. All the teams had been overwhelmed ever since the sliver of the sun had disappeared a month ago. It wasn't just the protests in front of the observatory.

Suicide rates had shot up. Scams had become more prevalent. The number of domestic violence cases was getting out of hand. Street vandalism was everywhere. Nobody had time to look for a woman who'd run away with her lover, but Dong Yiyao was unable to ignore the uneasy feeling that he'd had since the beginning of the case.

Despite being swamped with police work, Dong Yiyao still persisted through all these nights at Ma Gang's apartment. Tan Tan had not taken anything with her, not even her medical supplements. As a matter of fact, Ma Gang said, she had just purchased a new box of dried mugwort, which she must've planned on bringing to her doctor. Dong Yiyao couldn't say what it was that was nagging at him, only that something didn't add up.

A black-and-white bird landed on top of Dong Yiyao's car. Its eyes were a brilliant red. Dong Yiyao recognised it as a black heron. Careful not to disturb the bird, he reached into the glove compartment of his car and grabbed his camera. His daughters would be excited about this, he thought, they both loved birds. He quickly held the camera to his eye, but the bird flew off into the night before he could press the shutter button. With a disappointed grunt, he stuffed the camera into his pocket.

The window of the apartment went dark. Dong Yiyao looked at his watch. It was almost three in the morning.

Relieved that Ma Gang was finally going to get some sleep and realising that he was in dire need of some rest himself, Dong Yiyao only waited five minutes before stretching his legs and setting off to get that tobacco he'd been thinking about.

The following day was a public holiday – Dong Yiyao's first day off in weeks. As he walked, he started to think about the activities he could do with his daughters. Perhaps he could take them on a ride across town. His youngest always loved being driven around in a car. He could take them to the edge of Five Poems Lake, to the farm where his father used to work, play with the baby animals and watch the sun set over the desert.

Dong Yiyao believed that all crimes, in one way or another, happen because of people's ability to experience fear. This is the case on both an individual and a collective level. The former can kill a person. The latter can destroy a society. Many claim that courage can be gained through having faith, or love, or ideals, but to Dong Yiyao, it could only come from knowing the truth. That was why he'd become a police officer. That same resolve, he told himself, was propelling him to watch Ma Gang all these nights.

As much as he did not want to admit to it, he did have a more personal reason to probe further into Tan Tan's disappearance. The truth that he wanted to learn was bigger than this one case. The circumstances were much too similar to his wife's disappearance all those years ago. Just before she left, she had been talking about how she'd wanted to chop up the leftover green beans and add them to fried rice the following day. It was the only way to make Dong Ji eat vegetables, she'd said. She'd just been paid her salary, so she'd gone on a shopping day and returned with all sorts of things for the house: an orchid plant, a vase, a giant candle, a new cutting board. Dong Yiyao had always taken pride in being a perceptive man. It made him different from other men, he believed. If there had been any signs of his wife wanting to leave, he would've noticed them.

Yet he missed something. Dong Yiyao learned that a void was not an empty space, after all. Despite all the new things his wife had bought, without her, the house felt like it'd been hollowed. The only explanation he could offer Dong Ji was that her mother had left Five Poems Lake. He couldn't even say whether she was alive or dead. All these years, the weight of the unknown had been steadily chipping away at his willpower. He was exhausted, fed up with his incompetence.

The bell hanging in the convenience store tinkled as Dong Yiyao pushed the door open. The clerk was a young woman – a student, most likely. She was sitting on the stool behind the counter, drawing in a sketchbook.

Dong Yiyao grabbed a bottle of beer from the fridge and walked up to the counter for his tobacco.

'Do you mind rolling me one?' the store clerk asked

as she handed him the bag. 'I quit smoking, but sometimes I still want one.'

'Then you haven't really quit, have you?'

Dong Yiyao ran his trimmed nails over the sticker, struggling to open it.

'The rule is that I can't buy any tobacco,' the clerk said.

'When was the last time you had one?' Dong Yiyao asked.

'Two . . . maybe three days ago. I don't remember. Give it here. I have nails.'

He handed the bag over. She scratched the sticker with her blue nails.

Dong Yiyao looked down at her sketchbook. The page was filled with drawings of hands.

'That's a lot of hands,' Dong Yiyao said.

'I like them,' she said, peeling off the tape and throwing it into the garbage bin next to her. 'Here you go.'

'Can you draw mine?' Dong Yiyao asked.

'Why?'

'You must be bored.'

'Yeah, well, I am, but don't you have something better to do?'

Dong Yiyao placed his palm on the counter, fingers stretched out. After a moment of hesitation, the young woman opened her sketchbook and started drawing.

'Do you study art?' Dong Yiyao asked.

She nodded. 'Sculpture,' she said.

'What kind of sculptures? Hand sculptures?'

'You have a lot of questions.'

'It's the nature of my job.'

'What are you, a therapist?'

'Not quite, but in a way, I suppose I am.'

'I have no idea what that means,' she said, laughing.

'Lots of people come to me with their problems.'

She didn't ask any more questions and picked up her pencil. She sketched speedily and skilfully. It must've only taken her a few minutes to finish. She tore out the page and handed it to Dong Yiyao.

'You have to sign it,' Dong Yiyao said. 'In case you become famous one day.'

Wanting to brush off the comment but not very good at hiding her sense of achievement, the young woman signed the bottom of the page.

'Do you still want that cigarette?' Dong Yiyao asked.

'I guess not.'

'When you want to smoke, just give yourself a five-minute task. Your cravings will go away. Trust me, I've quit smoking many times.'

She laughed again.

'What's so funny?' Dong Yiyao asked.

The young woman waved her hand in the air and said it was nothing.

Dong Yiyao was carefully folding the sketch into a square when he looked out the glass door and spotted Ma Gang's scraggy figure on the other side of the street.

In a rush, he said goodbye to the woman and stuffed the drawing and tobacco in his pocket. He darted out of the store and across the road, hastening his footsteps but still keeping himself at a distance so that Ma Gang wouldn't notice. He'd forgotten to buy rolling paper.

Ma Gang was wearing the same leather sandals as always and they kept falling off his feet. With one arm, he was hugging a backpack close to his chest. The other one swung awkwardly with each rushed step. Dong Yiyao followed him until he stopped at the lake.

The water was still, as though it'd settled into a solid object. Lights could be seen from the buildings across the lake, but their stretch of shore was illuminated only by moonlight. A large bird flew over their heads – an owl, most likely.

Dong Yiyao watched as Ma Gang walked all the way to the edge of the water. He stood and stared at the still surface like he was debating whether or not to jump. Dong Yiyao observed him carefully, readying himself to go after Ma Gang if he really did decide to do something so cowardly.

After a while, Ma Gang plopped down onto the embankment. Dong Yiyao breathed a sigh of relief. He was glad to see that Ma Gang was not an idiot. Ma Gang sat immobile for a bit longer until he unzipped his backpack and pulled out an object that was too dark to be identified from where Dong Yiyao was standing.

Once Dong Yiyao moved into a position to get a

better view, he saw that Ma Gang was holding a metal watering can. He was also able to see that Ma Gang was crying. Dong Yiyao scratched at a mosquito bite on his arm as he walked up behind Ma Gang.

'Crying again?' he asked.

Startled, Ma Gang looked up at Dong Yiyao through his sunken, wet eyes.

'Who are you?' he asked, snot smeared on his nose and drool dripping from his bottom lip.

Dong Yiyao took his police badge out of his pocket and showed it to him. Still, Ma Gang wore a blank look, as though it meant nothing.

'I'm the one who came and inspected your apartment,' he said. 'After your wife disappeared.'

'My wife . . . my wife . . .'

That made him cry even harder.

'Come on,' Dong Yiyao said. 'Take a look at yourself. Aren't you embarrassed?'

Dong Yiyao pointed at the watering can. He'd had enough of Ma Gang's sobbing.

'What is that?' he asked. 'What are you doing in the middle of the night with a watering can?'

Ma Gang looked away. His crying subsided a little and turned into quiet sniffling, but then he must've remembered something and started choking on tears again.

'What do you want from me?' Ma Gang asked.

'I want you to go home,' Dong Yiyao said. 'And to stop crying by the lake like some sort of ghost.'

'Why do you care?' Ma Gang asked.

'You think I want to be here? I'm here because . . .' Dong Yiyao was running out of patience.

'Forget it,' he said.

He squatted down to meet Ma Gang at eye level.

Ma Gang's whimper escalated into hysterical bawling, not unlike the time at the police station when he told everyone that Tan Tan had run away.

'Are you trying to cry us another lake?' Dong Yiyao asked as he squeezed Ma Gang's arm and tried to help the man up. 'Here, let me take you home. I'll hold your backpack. It's almost four in the morning.'

Ma Gang shook Dong Yiyao's hand away and stood up all by himself. He was stronger than his appearance suggested. With all his might, he hurled the watering can into the lake with a grunt. It did not land far and sank within seconds, disturbing the surface for a moment as the water splashed and rippled outwards. It wasn't long before the lake found its way back to serenity again.

'I did it,' Ma Gang said. 'I did it! That woman won't get anything from me!'

Ma Gang's exhilaration was fiery red on his face. It was extraordinary, Dong Yiyao thought, for his expression to change from heartache to determination and then to elation all within seconds.

'What was in the can?' Dong Yiyao asked. 'Here, why don't you calm down and start from the beginning.'

'The beginning? The part where I met Tan Tan or the part where we got married? Or the part where I found her in bed with that brute of a man? Can you believe it? She was foolish enough to bring him into my house, fuck him in our bed, thinking that I wouldn't come home and catch them. Actually, you know what? I bet it's what turns them on.'

'And then what happened?'

'And then?' He snickered. 'You wouldn't believe it. She stood up straight, naked and all, and looked me right in the eye and told me that he understood her, as if implying that I didn't. But I saw right through her, you know? I knew everything she was thinking. If that isn't understanding, I don't know what is. Then she said she was lonely.'

He pounded his chest with a fist.

'I am the lonely one!' he yelled. 'Not her! I am! I have nothing left.'

Dong Yiyao found that a part of himself envied Ma Gang's ability to blame his wife for everything and cry like a child to a stranger. He could never do that.

'She married me because of that,' Ma Gang said, pointing to the spot where the watering can had sunk. 'My money. Can you believe it? She even admitted it to my face.'

'Are you out of your mind?' Dong Yiyao asked. 'That was money?'

'It was the one thing I had left that she wanted. My

only bit of dignity. Now that I've got rid of it, let's see if anybody else wants to be with me.'

He looked like he was about to start crying again, but something bright began to grow between his lips. For a second, before the light spilt out of his mouth and enveloped his head, Dong Yiyao saw dread in Ma Gang's eyes, like a goat about to be slaughtered; like Ma Gang knew that something horrible was happening to him and yet he could not even come close to understanding what it was. Though whether he understood anything did not make a difference. He was helpless to suppress the light. Instantly, Dong Yiyao's vision went white as Ma Gang vanished into the brightness.

Dong Yiyao's first instinct was to save himself. He stumbled backwards and rubbed his eyes, trying to regain his vision. Unable to look directly at the light, Dong Yiyao took out the camera from his pocket and aimed it at Ma Gang, snapping a photograph. The bright camera flash must've startled Ma Gang as Dong Yiyao saw him fall backwards into the lake. The light that came out of Ma Gang's mouth, as it sank, was gradually extinguished by the dark waters.

Dong Yiyao was able to see again. He did not know how long he sat on the ground, but when he finally came back to his senses, the chipped sun had just risen above the horizon and he was the only one left by the lake.

5

The older officer's toupee was lopsided and threatened to slide off his scalp at any moment, exposing his bare head. Dong Yiyao followed behind as the officer spoke in a gravelly, monotonous voice, turning his head between every few words to make sure he was being heard properly.

Dong Yiyao needed to know what exactly it was that'd happened to Ma Gang, but he hadn't been sure where to begin, so after some deliberation, he decided that he'd start by asking around about similar cases regarding missing persons. Reporting the events to the Commissioner had been out of the question. All the paperwork would've delayed the investigation, and it was entirely possible that they would have given the case to somebody else. Dong Yiyao had to do this alone. It was his own burden to bear.

He soon found out that there had been two reports of missing persons: one was Ma Gang's wife, and the other was a student named Liu Mu, from the Western

District. It wasn't entirely uncommon for people to disappear. Every few months, there were citizens who tried to leave Five Poems Lake, but those cases were easy to detect. They'd pack their bags, their valuables, food and water. This case was different. Just like Tan Tan and Dong Yiyao's wife, the girl had not shown any signs of having planned to leave. She'd even completed the assignment that was due the following week. This connection seemed rather far-fetched, he admitted, but his intuition told him that there was something linking these cases together. *What a stroke of luck*, he thought. He had not imagined that he would find a lead so soon. When he found out about this case, he didn't waste a single second before heading to the Western District Police Station.

'I trust you,' the older officer said as he turned a corner. 'Even though this is the first time we've met. But you know, I've heard about you. If I didn't know your excellent reputation, I would've thought that you had something to do with this. We just got the report this morning about Liu Mu and then you knock on my office door, not even an hour later, asking about incidents involving missing persons? That doesn't strike me as a coincidence.'

'It's not a coincidence,' Dong Yiyao said.

'Oh?' The officer stopped his footsteps abruptly. His toupee bounced a little. 'I knew it.'

'Liu Mu was seen in the Eastern District last night,'

Dong Yiyao lied. The confidence in his voice surprised even himself. 'The owner of the skewer stand told me that he saw her at around three in the morning on Tuesday. That means she was in my district just a few hours ago. You don't have all the information.'

'Oh?' The officer started walking again, slower this time. 'But her mother told me that she was last seen at the University.'

'From what I know, she went to the Eastern District in the middle of the night. To get skewers. When did she go missing?'

'She didn't attend her morning lecture. The professor called her mother and her mother came to me. Apparently, she had missed the previous day of classes too.'

Fortunately for Dong Yiyao, the officer didn't seem like he placed too much value on keeping victim information confidential. Dong Yiyao suspected that not many of the old officer's colleagues enjoyed listening to him go on about whatever it was that he enjoyed discussing. So far, he'd already told Dong Yiyao about his delayed retirement, his two sons who were both working at the television station, his favourite place to eat marinated goose webs, his daughter who was a beauty but growing older every day and incapable of developing a boyfriend into a husband.

He had also told Dong Yiyao that Liu Mu was a first-year student at the University. According to her mother, she was a wonder child with a bright future ahead of her. She had won numerous mathematics competitions, skipped an entire year in school, and received a full-ride scholarship to the University. There was no reason she'd abandon her studies. The only stain to note was that she'd been tormented by an illness ever since she was a child – something with her spine – but she'd seen a doctor only two weeks ago and there were no signs to indicate that it had worsened.

'I just thought I should tell you that she was in the Eastern District,' Dong Yiyao said.

'Thanks for the tip,' the old officer said as they arrived at the main gate of the Southern District Police Station. 'I'll call you when I have more information.'

'To be frank,' Dong Yiyao said, 'I want to have this case transferred to me.'

'I don't know if I can do that. The mother filed the report here.'

Dong Yiyao had to improvise, fabricate a reason for him to take over the case, even though it didn't make sense. Having seen Ma Gang conjure up the bright light from within his small, frail body, only for it to be extinguished by the lake a few seconds later, Dong Yiyao felt ashamed, for the first time in his life, for being a police officer. He was there. He could've saved him. He couldn't fathom why his instincts had betrayed him so readily. The feeling that left him paralysed had been nothing he'd experienced before. Even when the sliver of the sun vanished, he had not felt that way. He stood next to the lake that morning and made a promise to find out just what had happened.

The old officer scratched at his nose. A few strands of hair stuck out from his nostrils.

'I don't know,' he mumbled and sighed. 'I can't just give this case away. What should I tell the mother?'

'What you should tell her is that you received more information from the Eastern District and her daughter's case is now in good hands.' Dong Yiyao attached a firmness to his tone that was intended to push the old officer to overcome his indecisiveness.

'The hell with it,' the officer conceded, rather easily. 'I do have to admit that we are quite busy with all the protests. But you have to fill out the Case Transfer Request, got it? Bring it to me afterwards. We both need to sign it. We have to follow the proper procedure, for my sake and yours as well.' With a straightened back and arms crossed, the old officer looked at Dong Yiyao for assurance.

'Of course,' Dong Yiyao said.

He couldn't turn back now. He'd never before defied his responsibilities as a police officer, let alone falsified witness information. Anyone would've deemed him out of his mind for risking so much for people he hardly knew. What was more unforgivable was that he was willing to do so at the expense of an innocent, worried mother and a young woman who could be in danger.

A seed of shame was planted in the recesses of Dong Yiyao's mind that day. It would propagate quickly, roots reaching out to his limbs, tearing down the foundation of his conviction and making him question whether his blood was still pure or if it had become tainted with something sinful, whether his bones were white or black.

Liu Mu's mother pulled some tissues out of a pack and wiped the beads of sweat that had been sliding down her forehead. The sweltering summer heat was unforgiving.

'Skewers?' she asked and blew her nose with the same tissue. 'Amu never eats skewers. I want to talk to the skewer-stand owner. I know my daughter, I can help you find her.'

Dong Yiyao told her that because the skewer-stand owner was classified as a witness, his identity couldn't be revealed to anyone.

'But I'm not anyone!' she said. 'I'm her mother!'

'I promise you that I'll give you an answer within the next week,' Dong Yiyao said. 'I won't sleep until I find out what has happened to Liu Mu. To Amu.'

He meant what he said. He told himself that even if this had nothing to do with what he was looking for, at least he might be able to save someone. He was certainly more capable than the officer with the toupee.

'I don't understand why Amu would be eating skewers at three in the morning, let alone all the way in the Eastern District. I've always told her not to go out in the middle of the night by herself.'

Dong Yiyao sat down on the bench next to her. She had come to look for him at the police station, but because everyone else had yet to learn of this case, he'd made up an excuse and brought her to the pier. The paperwork usually took a few days to be processed, and in most cases, it was only for the sake of formality and would've been stored away in a filing cabinet before anyone could be informed about it. So long as the old officer didn't go probing around, Dong Yiyao should be able to work on this case in private.

'Now tell me about your daughter,' Dong Yiyao asked. 'Personality, hobbies, friends. Anything you can think of.'

'She's a smart girl,' Amu's mother said. 'Really smart. She's just not so good with people.'

She unfurled another tissue out of the pack and wiped her neck.

'She doesn't have many friends?' Dong Yiyao asked.

'I don't think she is close with any of her friends.'

'Why is that?'

'She's just always been better than them. On top of that, she's younger than all the others in her year.'

'Better than them? You mean in school?'

'Not just school. When she was little, maybe around

eight years old, she met another girl who played the flute. So Amu came home and insisted that she wanted to learn to play the flute as well. She tried so hard to surpass the other girl and she really managed to do it, even though she started learning at a much later age. She became much better. After that, they weren't friends any more. I asked her why but she wouldn't tell me. I think the other girl must've been jealous.'

Her worried expression momentarily gave way to a sweet nostalgia. With a voice filled with pride, she added, 'Amu is just destined for greatness.'

Dong Yiyao forced a nod. He didn't know what greatness was, whether it meant to be successful, to help others, to achieve fame, or to be happy. If such things could be predetermined, then the world would be a more twisted place than he could ever imagine.

'Did she do anything out of the ordinary before she went missing?' he asked.

It was a subtle shift, but Amu's mother began to seem hesitant and embarrassed. It was an expression that Dong Yiyao had seen many times in his career.

'Not really,' she said.

'If you don't tell me what you know,' Dong Yiyao said, 'you're only delaying the investigation, putting your daughter in more danger. I should also warn you that it's an offence to withhold information from the police.'

'Just let me talk to the owner of the skewer—'

'Forget about the skewer stand!'

Dong Yiyao stood up, his wide shoulders towering over the mother. She stopped speaking and looked down at her wool leggings.

'It's nothing,' she mumbled. 'Really, it's nothing. It's just that she got a bad grade on an essay.'

Dong Yiyao sat back down.

'That must've been hard for her,' he said.

'She's a strong girl,' the mother said, as though she felt the need to defend Amu. 'She's dealt with that spine of hers all her life. If she can live with that, I can't imagine a bad grade doing this to her. It must've been something before the essay. Something must've been wrong, otherwise she wouldn't have got a bad grade in the first place.'

'Maybe Amu ran away because she was afraid that you'd be upset.'

'Ran away? To where? You don't think she's left Five Poems Lake, do you?'

She grabbed on to Dong Yiyao's wrist like it was her lifeline. Her hand was small. It reminded him of the grasp of a baby.

'Running away is not as easy as you think,' Dong Yiyao reassured her. 'People talk about leaving all the time, but very few actually do.'

'She doesn't know how to cook. She won't ever make it.'

The woman's long curly hair fell in front of her face as she lowered her head.

'My wife played the flute as well,' Dong Yiyao said. 'She was a flute teacher, in fact.'

'Amu stopped playing years ago to focus on school.'

'That's a shame.'

The following afternoon, by the time the Eastern District Police Station received the phone call, Amu had already been admitted to the hospital. She was discovered by a retired man who was taking a stroll in the park with his bamboo bird cage. When he found her, she was unconscious behind a tree. The doctors diagnosed her with overexertion, qi stagnation and dehydration. The news of Amu's discovery was first delivered to the University. After travelling through a few rounds of calls, it inevitably made its way to the ears of the Eastern District Commissioner.

Dong Yiyao did not find out about Amu's hospitalisation until he was summoned to the Commissioner's office.

'You don't have an explanation?' the Commissioner asked, his face flushed red. 'Imparting false information? Are you out of your mind? This is not our case.'

'I need to be on this one,' Dong Yiyao said.

'Your needs don't matter,' the Commissioner said. 'Is it worth losing your job over? Or going to jail? Think about your children, for heaven's sake, Yiyao! You should know better! You should know that this is serious.'

The Commissioner plopped down onto his chair and gestured, with some impatient pointing, for Dong Yiyao to sit down across from him.

Rather than complying, Dong Yiyao straightened his back and stood even taller.

'Sit down!' The Commissioner's deafening command filled the office.

'I owe it to someone,' Dong Yiyao said, his chin held high. 'I made a promise.'

'Who?'

Dong Yiyao clenched his jaw. He could not tell anyone about what had happened to Ma Gang. Not before he had unravelled and straightened out all the threads. Luckily, it seemed that nobody had reported the man missing just yet. Maybe it was because he didn't have many friends. Even though Dong Yiyao hadn't got a clue what the light was and where it'd been born from, he sensed that it came with a danger, something that must be handled with caution. It killed Ma Gang, after all. If others were to hear about Ma Gang's case, the truth would become lost in all the commotion that would ensue before he could ever grasp it.

The Commissioner, confronted with Dong Yiyao's disobedience, slammed his desk, knocking over a container of pens.

'I'm trying to help you,' the Commissioner said. 'And you're as stubborn as a donkey.'

'Help is not what I need,' Dong Yiyao said.

'The police station is not your toilet, where you can come and foul and leave as you please. We have rules, ethics, order. Things that I thought you knew better than anyone else.'

'I knew very well what the consequences were,' Dong Yiyao said, 'yet I still did what I did. I must honour this promise, otherwise I would not be worthy of being here.'

'You're putting me in a tough spot,' the Commissioner said. 'You're my best officer.'

'I am undeserving of your praise,' Dong Yiyao said as he saluted his mentor. He took his badge out of his pocket and placed it on the Commissioner's desk.

'What is the meaning of this?' the Commissioner asked. All of a sudden, his tone had become unsure.

'It has been an honour serving the Eastern District Police Force,' Dong Yiyao said.

The Commissioner must've still been in a state of disbelief; he did not react. Dong Yiyao turned around and exited the room.

Just like that, Dong Yiyao threw away everything he'd worked for in his life. From the moment he had the case transferred, he'd known that this day would come, but a small part of him had wished for luck to be on his side anyway. As he walked down the corridors and the stairs, he struggled to suppress his pride and the feeling that he was wronged somehow. He felt at once a victim and a perpetrator. It was the justice he deserved, he repeated to himself: he'd done something horrible.

He needed to know that it wasn't cowardice that'd got in the way of him saving Ma Gang's life. There must've been something else, something out of his control.

He headed straight towards the hospital and found Amu's mother sitting next to her daughter's bed, begging Amu to finish the ceramic bowl of soup in her hands. Amu was a thin girl, with skin light like paper and a nose so small that it seemed like it could balance on the tip of a needle. Her eyes were only halfway open. It was as though a layer of film was covering her face, isolating her just a little from the rest of the world.

Upon seeing Dong Yiyao, Amu's mother put down the bowl of soup and stood up.

'Leave,' she said firmly. 'How could you lie to me? I thought you people represented justice.'

Dong Yiyao lowered his head and turned around. He wasn't even sure why he'd come. Perhaps he wanted to hear from Amu herself why she'd disappeared, despite the fact that in all likelihood the answer would have been disappointing. He decided he'd come back again when the mother was gone. Amu's eyes, which were looking vacantly at the television on the wall, drifted towards Dong Yiyao's direction, though it seemed that her gaze slid past him to some space in the distance.

Dong Yiyao waited in his car outside the hospital.

He'd forgotten to turn in his keys when he left the station. Patients and their families came and went, some with happier faces than others. He opened the envelope on the passenger's seat and took out the photographs that he'd got developed. At the very top of the stack was the one showing Ma Gang's body, illuminated white by the flash. His head was no longer visible, replaced by the bright light that'd come out of his mouth. If it wasn't for this photograph, Dong Yiyao might have thought that he'd imagined it all. How was he going to proceed from here?

Dong Yiyao studied the light in the photograph. It had seemed like it simply manifested inside Ma Gang and had expanded and risen until it could not fit any more. He put the photograph away and leaned his forehead against the cool car window, his breath fogging up the glass as he breathed steadily. He thought about what Ma Gang would've done, had he been able to save him from falling into the lake. Would Ma Gang have gone home? Would he have harmed anyone? If left to his own devices, would he have turned back to normal? Could it be that he'd already died the moment the light came out of him?

He stuffed the photos back into the envelope and stored it inside the glove compartment.

Amu's mother didn't leave the hospital until sundown. Dong Yiyao waited until the nurse at the reception counter had gone on her toilet break before slipping past the corridor and towards Amu's room. He knocked quietly and heard a girl's voice telling him to enter.

'You're the man from earlier,' Amu said, shifting uncomfortably in her bed.

'I'm a police officer,' he said out of habit, but did not correct himself.

'Why did you say that thing about the skewer stand to my mother?'

Dong Yiyao sat on a stool by the door.

'I made a promise,' he said. 'To someone who isn't here any more.'

'A promise? What's that got to do with me?'

'That's what I want to find out.'

Amu slipped a rubber band off her wrist and tied her hair into a loose ponytail. Under the night light, her figure was as elusive as a shadow. Dong Yiyao rolled the stool towards the bed.

'Where were you these past two days?' Dong Yiyao asked.

Amu shrugged.

'It's hot in here,' she said, fanning herself.

'What is the last thing you remember doing?' he asked.

'I want some air. It's hot.'

Dong Yiyao got up from the stool and pulled on the string that was attached to the ceiling fan. The blades began to spin slowly and gradually sped up until they settled into a rhythm, making a steady, creaking noise. A breeze swirled around the room and lifted up some papers on the nightstand.

'Can you help me get out of here?' Amu said. 'I want to get out.'

'You can leave when you're healthy.'

'I'm not talking about the hospital. I want to get out of Five Poems Lake.'

'So you were trying to leave?' Dong Yiyao asked. 'Then why didn't you bring anything with you?'

Amu bit her lip and averted her eyes.

'I wasn't,' she said. 'Not this time.'

'Then why do you want to leave all of a sudden?'

'I just don't want to spend the rest of my life here,' she said. 'And it's not an impulsive decision. I've been thinking a lot about it. It's just that this time, I really wasn't planning to leave.'

'I can't help you with that. Think about your mother.'

'Why should I? She doesn't think about me.'

'Of course she does. You should've seen how worried she was.'

'You don't understand. Even when she's caring for me, she's actually thinking about herself. If you're not going to help me, then just leave.'

Dong Yiyao looked down at the rug on the floor.

A corner had been turned up. He flattened it with his boot.

'Sorry to ask you this out of the blue,' he said, 'but have you seen or experienced anything strange recently?'

'Strange? What is that supposed to mean?'

Dong Yiyao wasn't sure how to explain himself. He should've known that this was a waste of time. Here was the case of a teenager trying to escape Five Poems Lake, a matter that had nothing to do with Ma Gang or the strange light. He wondered whether he might have been too desperate to find a link between the two cases. Maybe he just needed sleep.

Feeling defeated, he was about to leave Amu alone when she said, 'I've been feeling a little off, if that's what you mean.'

'Off?'

'My spine has been worse than ever. I can't keep my back straight any more. I can barely walk properly.'

'Oh,' he said. 'Have you told your doctor?'

'I don't want my mother to know. She'd freak out.'

'Still,' he said. 'You should tell your doctor.'

He knew that it wasn't his place to tell Amu what to do. She wasn't his daughter, after all. As far as she was concerned, he was just a police officer, which wasn't even true any more. He had no authority over her.

'Try and get some sleep,' he said as he opened the door. 'I'm sorry to have disturbed you at this hour.'

'I've decided,' Amu said. 'I'm leaving tonight. Whether you help me or not.'

'It's your life. Just remember that you only have one. Don't be stupid.'

He closed the door behind him.

6

Some hours later, when the night had fallen to its deepest, Amu tiptoed out of her room, dressed in baggy denim trousers and a black T-shirt. She carried a light backpack and winced whenever she tried to keep her spine straight. Dong Yiyao was sitting in the waiting area.

When Amu saw Dong Yiyao, she stopped for a moment and shot him a displeased look before walking past him.

'You're going to die out there,' he said as he stood up to follow her. 'Look at you. You look like you're eighty years old.'

Dong Yiyao could not turn away and let her walk to her death. He could not stomach the thought of failing to save another person. The moment he'd taken over her case, he'd become responsible for her wellbeing.

She didn't say a word. Dong Yiyao considered phoning Amu's mother, but if he did, she likely would prohibit him from ever talking to her daughter again. He couldn't blame her, but he'd come so far and sacrificed so much. A part of him still hoped to discover something else that he'd missed. Amu hadn't told anybody what she'd done in the past two days. There were still blanks that needed to be filled. She'd confided in him, shared with him her plans to venture senselessly into the desert.

He followed her all the way to the bus station. She dragged herself along with much pain. Not once did she look back at him.

They sat side by side on the empty plastic bench. Sweat dripped from Amu's hair onto her face, which now reminded Dong Yiyao of the girl's mother. Earlier, he had thought that they bore no resemblance to each other. Her mother had a round face, protruding front teeth and lush hair that bounced up and down while she talked. Amu was thinner, more withdrawn, more tired. When it came to the eyes, it was Amu's that were more motherly. But now, observing Amu from this new angle, there was no doubt that she was her mother's daughter. Dong Yiyao couldn't say what it was that brought forth the similarities all of a sudden – it wasn't that the two women had changed in appearance. Maybe it was something that normally flowed in the blood and, only spontaneously, at certain angles, would reveal itself on the outside.

'I just want to *be something*,' Amu said finally. Her breath was laboured and her voice came out like a puff of smoke. 'Something,' she repeated. 'Someone.'

'Like what?' Dong Yiyao asked.

'I don't know. Something that defines who I am, how I'm different from the others. I want an identity that only belongs to me.'

'I can't say that something like that really exists,' Dong Yiyao said. 'We're all more similar than we'd like to believe.'

'Then what's the point? What's the point of all of this?'

'Being similar to others isn't as bad as you think,' Dong Yiyao said.

'You won't get it,' she said. 'I've been told my whole life that I'm not like the others, but I can't seem to figure out how it is I'm different. I'm scared that when I die, I'll be gone for ever. There will be nothing left of me. I want to live for ever. I really do, officer. But my body can't do that, so I want to find another way to get as close to eternal life as I can.'

'You're still young,' he said. 'You don't have to think about all of that yet. But if you go out there, you'll probably die sooner than you think.'

'Don't talk to me like I'm a child.'

The bus emerged from the corner and made a screeching noise as it slowed to a stop. Its once white exterior was now largely cracked and rusted. One of its headlights was flickering and the door was scarcely hanging on. Every year, more buses broke down beyond repair. Over half of the bus stops around Five Poems

Lake were now terminated, turned into moss-covered booths where teenagers gathered to smoke and drink in the afternoons as their days rotted away.

Behind the windscreen was a piece of paper with its final destination – somewhere near the northern farms – written on it with a black marker. Two passengers alighted from the back while the bus driver reached over and pushed open the front door. Amu hauled herself up the steps.

'I'll take you in my car,' Dong Yiyao blurted out right as Amu was about to give a coin to the driver. 'I'll drive you to the desert. To wherever you want to go.'

She hesitated and withdrew her hand. She looked at the bus driver, who was distracted by something in his rearview mirror.

'You can't stop me,' Amu said.

'My wife disappeared,' Dong Yiyao said. 'She disappeared, just like you did, taking nothing with her. Only she never showed up fainted behind a tree. She never showed up anywhere.'

A passenger yelled from his seat and the bus driver, annoyed at being rushed, thrust his frustration upon Amu and asked, in a mean-spirited tone, whether she was getting on.

'There are some answers I need,' Dong Yiyao said. 'If you help me, I'll help you.'

Amu turned around and stepped back onto the pavement. She stuffed the coin back into her pocket.

The driver closed the door and shook his head vigorously at her before driving off.

Dong Yiyao led Amu to his car. They stopped by a convenience store and bought some water, a bag of fried spring rolls and some canned food. Then Dong Yiyao stepped on the gas and drove north.

'I don't know where your wife is,' Amu said.

Sinking into the car seat, Amu became so small that she looked like a little girl. She stared out the window at the dark countryside road that was only dimly illuminated by the headlights. It felt as though they were driving on the same stretch of road over and over again.

Dong Yiyao reached into the back seat and grabbed two cans of orange soda. He handed one to Amu.

'It's not even cold,' Amu said.

'It's all I got.'

She pressed the can against her cheek.

'My mum is going to hate you,' she said.

'It sure seemed like she already does.'

In the corner of his eye, Dong Yiyao saw Amu hide her smirk.

'She doesn't let me drink soda,' she said. 'It's bad for the brain, apparently.'

Amu traced the rim of the can with her index finger. A cluster of houses came into view on the right side of the road. A few of the lights were on. As their car continued onwards, the lights grew ever more faint and small until they disappeared into the distance.

'I'm not expecting you to find my wife,' Dong Yiyao said. 'I just need you to tell me what happened, and where you were for two days.'

'I didn't go missing,' Amu said. 'They just didn't look for me hard enough.'

'What's that supposed to mean? I don't have time for jokes and riddles. There are enough mysteries for me to solve as is.'

'Believe what you want. I'm just holding up my end of the bargain. All I did was skip some lectures. Is that so hard to believe?'

She grabbed Dong Yiyao's soda, opened it for him and handed it back.

'My mum can't fathom the thought of me being in places other than where I usually am. It's like I'm a train. She can't imagine me being off track.'

'They teach you about trains at the University?'

'In my history course, yeah.'

'What do they know about trains? Do they tell you where the broken tracks once led to?'

'If they did, I could've just followed the map of the tracks instead of asking you for help.'

'Then why spend time teaching about an obsolete thing that nobody really knows anything about?'

She shrugged.

'Trains are cool, I guess.'

Dong Yiyao gulped down some soda.

'So have you always listened to your mum?' he asked.

Amu looked down and pulled at a loose thread on her T-shirt as though she was ashamed.

'So you were just walking around town for two days?' Dong Yiyao tried again.

'You know when you asked me if something strange had happened lately?' she said. 'I feel like my body is falling apart, like it doesn't want me to live, and precisely because of that, I'm starting to have this overwhelming desire to live for ever. To fight against it. I can't even control it, this desire, but I can feel it expanding inside my heart.'

'You're too young to be worrying so much about death. Your mother said that the doctors didn't seem too concerned about your condition.'

'You don't know me,' Amu said. 'You don't know how I feel. It started off as a thought. An exercise. I don't want to die here, as a nobody. I was counting up all the things I could do with my life . . . and then I realised that none of it would be mine. Whatever I become will somehow belong to my mum. I know this sounds stupid, but I feel like if I keep going on like this, I'll blow up. I need to find something that is mine, something that will survive for ever.'

'And you think you can find that out in the desert?'

'There's a house out there,' she said, pointing forward. 'In the desert to the north.'

'Did they teach you that in school too?'

'I saw it on a wall. Somebody wrote a whole story

about it with chalk. I admit I wasn't just wandering around town. I was scouting. Looking for the best way to escape.'

'Someone wrote something on a wall, and you believed it? I thought you were the smartest in your class.'

'It doesn't matter if it's real,' she said. 'I'm going anyway. I can't breathe in this town.'

She cracked open her own can of soda and took two sips. The night was so quiet that Dong Yiyao could hear the liquid fizzing in her mouth. He sighed, finally accepting the fact that Amu was just another girl fantasising about something in the world beyond Five Poems Lake. There had been many cases of teenagers like her trying to leave, thinking that the world was brighter out there. Their motivations usually had something to do with the decay of Five Poems Lake. They couldn't bear living in a place where deterioration happened at a faster rate than creation. They believed that it was up to them, the younger generation, to pave a way for themselves and find somewhere else, somewhere more alive.

Thirty or so minutes after they drove past the last farm, the dirt road ended and the sands began. Dong Yiyao turned off the engine. Darkness swallowed the car.

'This is as far as my car can go,' he said.

'Are you heading back?' Amu asked.

'I can't leave you here in the dark all by yourself.'

She made an effort to hide her relief, but Dong Yiyao detected it all the same.

'Let's wait until dawn then,' she said with a satisfied nod.

Some clouds cleared up and moonlight found its way into the car. Dong Yiyao rolled down his window and looked at the distant sky. The edge of the full moon was so crisp that it made the moon seem thin, like it was a circle cut out from paper. The cicadas' chirping was particularly loud that night.

At some point, they both fell asleep. Dong Yiyao woke again to the summer heat. The morning sun was strong. It looked as though it was still a full sphere in the sky, like the disappearance hadn't happened. In front of the car was a barren desert that blanketed the earth like static, orange waves. When he looked over at Amu, he saw that she was already awake and on her lap was the photograph of Ma Gang.

'Where did you get that?' Dong Yiyao snapped at her.

'It's not like you were hiding it very well,' she said.

'Who taught you to steal?' He seized the photograph from Amu. She surrendered it easily.

'What's going on in that photograph?' she said.

'Let's go back to town,' he said. 'We're done playing around.'

'I'm going out there. You're not going to change my mind.'

Dong Yiyao looked down at the image of Ma Gang. He didn't know whether he was looking at a dead man.

'This plan of yours isn't going to work,' Dong Yiyao said. 'Look at you, you can't even walk with your back straight. You'll die, and no one would even find your body. Isn't that the very thing you want to avoid?'

'You don't have to keep telling me that,' Amu said. 'I know. I'm not an idiot. But I'm still going.'

'I failed to save this man!' Dong Yiyao blurted out. He held the photograph to Amu's face and pointed at the centre. 'I watched him turn into this thing. I can't be responsible for you too.'

Now it was Amu who wore an expression of doubt, as though she was judging whether Dong Yiyao had lost his mind.

'The light came out of his mouth,' Dong Yiyao said. 'It was brighter and hotter than any light bulb or fire I'd ever seen. He fell into the lake before I could react and the light was gone in an instant. I can't make the same mistake again. I need to take you back. It's for your own good.'

'I don't know why you're so adamant about interfering with my life,' Amu said, looking at the photograph that he was holding up to her face, 'but for sure it's not for my good.'

She opened the door and pushed out of the car.

'There are people who love you,' Dong Yiyao said.

'That's not why you want to bring me back.'

Dong Yiyao did not respond. He could not even look her in the eye.

'Goodbye,' Amu said hesitantly. She drew out the last syllable as though it was a string connecting her to Dong Yiyao.

She stepped into the sand, her shoes sinking a little with every step. There was nobody else around. The last farm they'd passed was miles away. He gazed out at the desert until Amu's slight figure descended down a slope and vanished from sight.

Dong Yiyao could've followed her – he could've stopped her with force – but he had come to know just how stubborn she was. She needed to have tried to do it herself. Only then would she come to understand her limits and what she really wanted. He knew that she would come back. Even if she proved to be more persistent than he'd thought, with that frail body of hers, it wouldn't take him long to catch up with her.

He shut his eyes. The silhouette of Ma Gang with the light on his shoulders grew inside Dong Yiyao's consciousness and closed in on him. He tried to discern Ma Gang's face, but the image in his memory was like the sun in his eyes. For a moment, he thought he felt his skin grow hotter. He saw the watering can in Ma Gang's hands, the sandals that were falling off, the ends of Ma Gang's bones stretching out his skin.

Dong Yiyao opened his eyes just enough so that sunlight could be let in. He was exhausted. What drained him the most was the thought that he'd got nowhere with his investigation. He felt sorry for dragging Amu into all of this. Like she'd said, all she did was skip a few days of school. It was evident by now that Amu's case had nothing to do with Ma Gang, Tan Tan, or his wife. Even when she saw the photograph, he could tell that she did not have information regarding the light. All this time, he must've been too desperate for some sort of connecting thread. He'd thrown away his job, he hadn't gone home in days, he'd stolen a police vehicle, he'd let Amu venture into the desert alone, all for nothing.

But, if he hadn't waited at the hospital, he told himself, Amu would've really got herself in danger. At the very least, he would be there to save her life should she fail to return in a few hours. He wondered how many lives he had to save for Ma Gang to forgive him.

He let his eyelids fall heavy again. Once Amu had seen enough of what was out there, she would learn to give up on this pointless pursuit. She was irrational, he thought, but not stupid. At that point, he would drive her back, bid her goodbye, and from then onwards, he would leave her and her mother out of this mess.

BEACONS IN
FIVE POEMS LAKE

7

Snow fell like it was never going to stop. It'd been three days since Red Bean showed the video on TV and nobody seemed to have turned up with any real information regarding the Beacons. Red Bean hosted panel discussions every morning, bringing in guests from all walks of life, but it seemed like they were getting nowhere.

People had started to make things up. Ba used to say that rumours are birthed from the cradle left empty by the truth. In the case of the Beacons, the rumours fed on this void and became all the more rampant. Some people were convinced that the Beacons were sun worshippers who had made themselves glowing helmets. When the sun first faded, it was reported that a group of people used the phenomenon as evidence to show that the sun was a god and it wanted us to perish. It wasn't a popular opinion, and it wasn't until I overheard a man talking about it that I came to believe sun worshippers really existed.

Some of the more superstitious among us understood the Beacons to be a prophecy warning everyone about the impending end of the world. Others accused them of being the reason the sun was fading in the first place, saying that those who believed in the supernatural must've channelled some kind of illegitimate, evil power to bring us all towards our demise. It wasn't that no one was aware of the irony in that notion; I suspected that it was only because they didn't know what to believe in. People were not fools, they were just afraid.

There was no evidence to confirm that any of this speculation was true, though none had been proven wrong either. Most people, as I had observed, didn't even believe that the Beacons were real. Even I knew that it was too much to ask from them. Most of them had not seen them in person, after all. They dismissed the video as being fake, filmed with actors and props, made up by Red Bean to win back viewers.

Those who had seen the Beacons with their own eyes were still few and far between. The ones that wished to speak about what they'd witnessed most often sounded like they were telling a far-fetched tale that they themselves didn't even believe in. Others who didn't want the attention simply kept their mouths shut.

Dong Ji was by now adamantly against me filing that official police report about Driver Hua. I wasn't sure what had changed. She spent all her time trying to conjure up different plans regarding what to do with the photograph we'd found. Every time she came up with something, she would abandon the idea the same day.

At first, what I'd thought was a heavy sense of duty tugged at my heart. Later, I realised that the feeling was not duty, but rather the burden of knowledge, like a giant mountain pressing me down into the ground.

We were at the market hoping to find flour and vegetables when I decided to bring up the subject again.

'No,' Dong Ji said, not missing a beat.

'Why? Shouldn't we help the police?'

'Is Gao Shuang pressuring you again?'

I told her it was my idea.

We walked up to a vendor who had a line of small plastic baskets on her table, each carrying two heads of broccoli on sale for exorbitant prices. The broccoli didn't look attractive, but the days when we could select vegetables based on their appearance were long gone. Dong Ji picked up a basket.

'Are you crazy?' I said. 'It's broccoli, not gold.'

'You need it. Look at your skin. It's all dried up. There is no elasticity at all.'

'Who cares about my skin? We still have pickled cucumber at home.'

'I care.'

She handed the basket to the vendor, who weighed it, showed us the number on the scale, and then stuffed the heads of broccoli into a paper bag that was damp from the cold.

'You have to stay healthy,' Dong Ji said.

She handed some bills to the woman, who, even though we were giving her money, refused to interrupt her conversation with another customer.

'I can file an anonymous report,' I said quietly.

'I just don't want you getting caught up in all this,' Dong Ji said.

'I'm already caught up,' I pleaded, hoping for her to understand.

She wouldn't look at me and continued to browse the stands. I followed her.

'We need milk,' she said. 'Look for milk.'

She craned her head over the crowd and scanned the market. There was a young man selling milk in the corner. Behind him were shelves filled with condiments and dried foods. We squeezed over to him.

The young man was pale, with papery skin that hardly concealed the veins beneath. He stood up tall when we approached and turned so that we could browse the boxes of nuts, dehydrated mushrooms, various seeds and bottles of sauces.

'We want four cartons,' Dong Ji said.

'Of milk,' I added.

'Don't you want to buy some more?' the young man asked. His deep voice made him sound much older.

'Who knows how much longer the cows will be producing milk.'

'We have a small family,' Dong Ji said. 'Too much milk will go bad.'

'I'm sorry,' he said.

What a strange thing to apologise for, I thought. As the sun had continued to chip away throughout the years, I'd noticed that people had started to apologise more often.

'At least wait until I've figured something out about the photograph,' Dong Ji said.

She stacked the milk cartons sideways and lowered them into my bag.

'Wouldn't filing the report get us closer to finding out what's going on in that photograph?' I asked.

'People are going to come asking you questions,' she said. 'Everyone will know. What if the Beacons come and try to hurt you?'

'They seemed peaceful in the video.'

'What about Driver Hua? Remember what he tried to do to you?'

'If the Beacons are going to harm people, then shouldn't I warn the police?'

'How come you have to be like this? Just give me some time, all right? I'll figure something out.'

How come you have to be like this? was Dong Ji's favourite string of words when she was frustrated with me. Most of the time, I'd let it go and leave her alone. Occasionally, when my temper overtook my better judgement, I would push and our argument would escalate until, inevitably, it would hurt us. Today, I could tell that we were both tired.

'OK,' I said. 'I'll wait.'

That afternoon, after some rumination, I paid a visit to Driver Hua's wife. I had an idea that, if I ultimately were to file a police report, I should ask for her permission first. It was her life and reputation that Driver Hua cast aside and threw in the garbage with his actions, and I felt it would have been unfair for her not to have a say in how it played out.

I found Driver Hua's address in Yeye's phonebook and made my way over. His place wasn't too far from our house, a twenty-minute walk north, but with all the snow on the ground, it took me almost an entire hour. He lived in one of those low-rise apartment buildings at the edge of the Eastern District. Not long ago, one of the taller blocks nearby had collapsed in the middle of the night, killing half of its residents. The reason, according to official reports, was that ten makeshift units built on the rooftop had added too much weight for the already old structure to support. Driver Hua's story was that a woman was visiting her husband from the countryside and their lovemaking was so passionate that it shook the floor too vigorously. I asked him whether the extra units on the roof could've played a part, but he simply joked that passionate sex was more powerful than I could ever imagine.

Driver Hua's apartment was on the ground floor. The floor tiles were covered in sludgy footprints and much of the dark grey paint on the walls had peeled off. I found his door at the end of the corridor and knocked on it a couple of times. Nobody answered. I tried with more force and the door to the left of his creaked open. A plump woman with frizzy hair popped her head out.

'Who are you?' she asked.

'Does Driver Hua live here?' I asked.

The woman pushed her door all the way open. She was holding a bowl of rice topped with pickled garlic, and the steam from the rice rose up to her chin. Her mouth made an odd sucking sound, like she was kissing someone or salivating from a piece of sour candy.

'Not any more,' she said, angrily. 'He hasn't come back in a week. His rent was due two weeks ago.'

'What about his wife?' I asked.

'She moved back to the farms to help their son with school.'

'When did she move?'

She assessed me for a second. 'A little over a month ago,' she said. 'Who are you again?'

'I work with Driver Hua.'

That was the best answer I could come up with – the only connection I was willing to admit I had with that man. The woman didn't bother to hide the judgement and apprehension in her jaded eyes.

'Are you going to see her?' she asked.

'Yeah,' I said. 'I hope so.'

She did the sucking thing with her mouth a few times as she looked me over again.

'Wait here,' she said at last.

She went inside her apartment and closed her door. A few seconds later, she opened it again. The rice bowl in her hand had been replaced with a key.

'She left a bag over there,' the woman said. 'Take it to her. I was going to throw it out, but I have nothing against her. She's a good woman. Kept to herself most of the time.'

It took her a few tries to unlock Driver Hua's apartment. Her fingers were swollen at the joints and they seemed to have grown numb too from the cold. After some fidgeting, I heard the lock turn and followed her inside. The place couldn't really be called an apartment. It was just a room. If I stretched out my arms, I'd be able to touch the walls on both sides. There was a double bed pushed against the window, two plastic chairs facing one another and a wooden stool in between acting as a table. A rusted, portable gas stove sat atop a small refrigerator.

'Here,' the woman pointed to a duffel bag that was on the floor. 'Make sure you tell her I didn't open it.'

'Do you know her address?'

'You don't?'

I shook my head. She looked at me suspiciously again but must've decided that the bag was more troublesome than the possibility of my malicious intent. She went back into her apartment and scribbled the address down on a page from a notebook and ripped it out for me.

'Remind her,' she yelled towards me as I was leaving, 'about the rent!'

The bus to Driver Hua's village was unheated and screeched every time it made a stop. Most of the passengers were men. Some were sleeping. Those who were awake were smoking. Two women with wrinkled faces sat in the front. One of them wore a cherry-red coat and the other had on a purple quilted jacket: I thought they resembled dried prunes. They were munching on roasted sweet potatoes.

Most houses were still built from earth and stone this far out from the lake. When I got off the bus with the two women, I saw a stray dog with her puppies huddled together next to a small hardware shop. A teenage boy was gluing rectangular sheets of foam onto the walls of the shop to insulate against the brutal cold.

I asked the boy for directions.

'Over there,' he said in a laboured voice and pointed towards a hill.

'Can you still grow anything on these farms?' I asked, after I thanked him.

He shrugged as he wiped his nose with his sleeve.

'Some crops are still growing,' he said. 'But I don't know. They all say the frost is only going to get thicker around here. It's already killed the roots of most of the plants.'

'What are you going to do?' I asked.

He picked up a sheet of foam and ran a tape measure across it.

'We grow what still wants to grow, I guess. And kill the pigs. They eat too much, and their meat sells for a lot. Eggs, too. People like to buy eggs, and hens are cheaper to feed than cows and pigs. But a lot of the hens have died from the cold.'

Yeye worked on a farm once. It was only for a short period but, in his early twenties, he had no intention of inheriting the pharmacy, so he moved as far away as possible. Every morning, as he ploughed the soil, he would gaze towards the endless desert, contemplating the differences between a cage and this place he called home. He often had the impulse to just take a step into the sand and see what was out there, but time and again, the farm pulled him back. He told us that sowing those seeds must've planted him to this soil too.

I missed Yeye in that moment, and the boy must've noticed the sadness on my face, because he took a piece of candy out of his pocket and offered it to me.

'My little brother thinks that the sun is like a meat pie,' the boy said, 'and someone up there is taking bites from it. He thinks that they'll just cook up another one once this one is gone.'

'Maybe we should all learn from his attitude,' I said.

'He's an idiot.'

'Don't be so harsh,' I said. 'Siblings stay with you all your life.'

'I guess so. But idiots won't survive long around here.'

He pointed me towards the hill again and returned to his work. I wished him luck and went on my way.

There was a cluster of clay and stone houses on top of the hill. I found the Huas' home smack in the middle of them, so ordinary that I almost missed it. A red couplet, washed out and creased, was glued to either side of the door. The few characters on the bottom were smaller and squeezed together: whoever had written them must've miscalculated the length of the paper. I knocked a few times and waited. A woman called from behind the door.

'Coming! Coming! Who is it?'

She was there before I could answer. She was a little taller than me and, though she wasn't thin, she appeared rather flat, like somebody had stretched her out sideways.

'Are you Driver Hua's wife?' I asked.

'That's me,' she said with an indecipherable expression.

'May I come in?'

She looked at me, waiting for an elaboration.

'Oh,' I said. 'Here. This is your bag, I believe.'

'You came all the way for the bag? Did the landlady send you?'

'I own a pharmacy,' I said. 'Your husband used to deliver supplies for me. I just want to ask you some things about him.'

I stopped there. I wasn't quite sure how to explain myself. Maybe I shouldn't have made such an impulsive trip. I wondered why I hadn't prepared for this conversation on the way here.

'Let's talk inside,' she said and pulled open the door some more.

She led me through a small courtyard and into the house. There was a brazier in the middle of the room, which made it warmer inside, but not enough for us to take off our coats. She dragged a wooden chair closer to the fire and gestured for me to sit down.

'Do you want some hot water?' she asked.

'That would be nice. It's colder here than in the city.'

There was a bed by the inner wall. This room must've once seen plenty of sunlight: much of the dark wooden furniture had been bleached yellow. A plastic clock and a little mirror hung next to the window alongside some photographs of a boy I presumed to be Driver Hua's son. A wooden dresser painted red was standing opposite the bed with all sorts of items scattered on top – a bag of tobacco, papers, a standing calendar, pens, textbooks, bottles of half-finished drinks. There was also a black leather wallet. The buckle was in the shape of two golden birds facing away from each other.

She poured me some water from the kettle on top of the brazier and then perched against the dresser.

'I can't sit for too long,' she explained. 'It hurts my knees. Arthritis.'

'You have to keep your legs warm.'

She waved her hand, indicating to me that her ailment was way past that.

'Enough about my legs,' she said. 'So what did my husband do this time?'

'Has he been home recently?' I asked hesitantly.

'Did he try to do something to you?' she asked, ignoring my question.

My eyes shied away from hers. I hadn't expected her to ask so bluntly. Had I given some signal, said something to prompt her? No, I'd been careful.

'Son of a bitch,' she said.

Embarrassed, I brought the cup of hot water to my face and blew on it. The steam was warm and wet on my nose and cheeks.

'Shouldn't your son be back from school by now?' I asked.

'He lives with his grandmother on school days.'

'I thought he lived here.'

'He likes it there more.'

I drank some of the hot water; it burned my throat and chest as it travelled down.

'Driver Hua turned into a Beacon,' I said.

She rubbed her hands together and shook her head.

'Are you not surprised?' I asked.

'He hits me, you know?' she said. 'And I hit him, too. He thinks I have another man. I think he has other women. But let me tell you, woman to woman, I couldn't care less.'

'Do you?'

'Do I what?'

'Have another man?'

'Never. One is already too much for me. I just want to be by myself. I feel cleaner that way.'

'What about your son? Wouldn't it be easier to have someone help you raise him?'

She smiled and looked at the photographs by the window.

'He's a good kid. He doesn't really care for me though.' She pointed at one of the photographs. 'That's what he looks like now,' she said. 'The other photographs are from years ago.'

She bent over and began massaging her legs. 'I never wanted to get married,' she said. 'When I was young, I was afraid of marriage. I didn't want to count out chopsticks for dinner every night. I feared knowing exactly how much rice to cook for two people, and then for two people and a child, two children, three children . . . I didn't want to know that.'

'What changed?' I asked.

'Nothing,' she said. 'Like all young girls, I didn't really make any sense. As much as I didn't want to marry, I still dreamed of romance, sex and children. Before I had a chance to decide what I really wanted, I found myself in my wedding dress and a piss-drunk husband stripping it off.'

The words gushed out of her like she'd been waiting for ever to tell someone. She stopped herself and took a deep breath.

'He showed up at my house a week ago,' I said. 'At night. He was drunk, I think. A bright light, well, it pushed itself out of his mouth and his head just kind of disappeared into it.'

'So that's how it happens, huh?' she said. 'I never would've imagined that the light came from inside them.'

'Do you have any idea why it could've happened to Driver Hua?'

She gave me a look, as though she was sending the question back my way.

'If it really did happen,' she said, realising that I had no idea why, 'then he must've brought it upon himself.'

The fire was dying out. She walked into the kitchen. I heard her blow her nose. Soon, she was back with two more pieces of wood, which she added to the fire.

'You know, not once has he ever satisfied me?' she said. 'He always just assumed that what he liked to do in bed was what I liked.'

'Have you told him that?'

'There was no point.'

The flame in the brazier embraced the wood and found new life again. Outside, the setting sun slipped off the horizon.

'He used to say that he wanted to live for ever,' she said. 'What on earth did he have to live for? Even if somebody found a way to be immortal, he'd be the last in line.'

'Maybe his wish came true,' I said. 'The Beacons could be immortal, for all we know.'

'Well, like I said, I don't give a damn.'

'I want to file a police report,' I said. 'I just need your permission.'

She winced as she squatted down next to me and stretched her hands closer to the fire.

'I don't want our son to find out,' she said. 'He wasn't a good husband but, believe it or not, he was a good father.'

'He was a good truck driver too.'

I didn't want to praise him but, as much as his wife would've liked to say she despised him, I could still identify a delicate emotion beneath her words. I wouldn't call it affection or love, but something built up from years of marriage that was more sweet than bitter.

'I'll tell the police not to disclose his name,' I said.

She nodded.

I finished my water, which had grown cold by now

and taken on the metallic taste of tap water. After saying goodbye, I stood up and she asked me why I thought the sun was fading.

'I'm just a pharmacist,' I said. 'I will never know the answer to that.'

'I always thought the disappearing sun is our subconscious wishing for darkness and privacy.'

'There's no proof of that,' I said.

'Proof doesn't matter so much to me.'

'By the way,' I said. 'Driver Hua told me that you've made plans for the future when the sun is gone.'

'They were not real plans,' she said with a smile. 'Though they did make me feel better about all of this.'

'Do you really think our subconscious is powerful enough to influence the sun?'

'I don't know. It's just nice, you know, to have an answer for things. Makes life easier.'

'Even if it's not the right answer?'

'As long as it makes things easier while I'm alive. Also gives me something to tell my son.'

If I were Ba, I would've come up with a counter argument, but I wasn't him. I empathised with the need to rationalise things with stories, with dreams, with anger, with ideas that aren't necessarily correct.

By the time I managed to catch the minibus back, it was almost eight in the evening. Under the eerily bright moonlight, the bus made its way through the fields and towards the lake. A mother with a toddler sat next to me. The little girl watched my face as we drove. She didn't smile, nor did she cry. She just looked at me with her deep black eyes. During the entire ride, never once did her mother turn her head towards me. It made me wonder: at what point in our lives do we lose the curiosity of a child and learn, instead, to look away? And when will we ever think to look again? The mother and child alighted two stops before mine. When I finally made it home, Dong Ji was heating up some leftover sour cabbage and potato starch noodle stew for me, as though she'd anticipated my return. She asked me where I'd been and I answered with a lie.

The next morning, after Dong Ji left for work, I closed the shop and went to file the report.

Gao Shuang stayed with me the entire time, but it was a young woman who noted down the details. As I recounted the events of that night, the young woman looked me straight in the eye, as though with every word I said, she was trying to discern whether it was me who was not in my right mind or if the world that I was describing was not on its right axis. She was professional, though, and made sure not to miss anything I said. Gao Shuang listened with a clenched fist. From time to time, he squeezed my shoulder with his other hand to comfort me, but to my surprise, I felt no need for that. It wasn't that I'd forgiven Driver Hua, it was only that I didn't feel like a victim. Miss Pan was a victim. And there was Driver Hua's wife, too. Why did I turn out to be the only witness to his fate? Why not his wife? When is it that two people become husband and wife? Does it really happen on their wedding day? Or can it be some other time? And when do we come to know one another? Is it when we meet? Or does it occur when one begins to long for the other? Do our fates really intersect at the moments we think they do?

On my way home, I saw two women with wide brooms sweeping snow towards the sides of the road, piling it into walls the height of my shoulders. As they worked, snow continued to fall, but they kept on sweeping.

Only two customers came in that afternoon. The first was a woman who thought that too much internal heat had built up inside her husband's body. He had pimples on his face, she told me, and his breath reeked. Sometimes, he'd complain about his gums being in pain. At this rate, she said, she was sure that all the heat would turn him into a Beacon. I told her that he was most likely going to be all right. Internal heat was a concept that'd existed for a long time, and the Beacons were a more recent occurrence. It was unlikely that the two were related in any way, but she seemed frightened and unconvinced. I gave her the medicine and some dried roselles and did not debate with her any further.

The second customer was a man with two giant bags

under his eyes who whispered that he had a low sex drive. He said he wanted to have all the sex he could before the sun disappeared. I explained to him that he needed to exercise restraint, that any herbal medicine doctor would've told him that he had to take better care of his kidneys, especially if he ever wanted to have children. I sold him some qilin pills for his problem and sent him on his way.

Another night passed and Red Bean showed up at my door in the morning. The firm sound of her knuckles on the wood reverberated in the streets, which had been as quiet as a cemetery the entire morning. A few more Beacons had popped up in the Northern District the night prior. Red Bean reported that they'd been sighted near the wood workshop where Yeye brought us to get Ba's altar made. According to those who lived nearby, the Beacons had been wandering around that area all night. A woman told Red Bean that she didn't even dare to look at them, afraid that she'd be cursed somehow. Her neighbour had more courage, it appeared, and noted that the Beacons acted as though they were searching for something, somewhere to go – a purpose. Their faces were gone, of course, but a subtle sense of purpose could be identified from the way they walked, turned their bodies, tilted their suns.

People hid at home with hammers and sticks by their doors and watched the streets through their windows.

Their anticipation contained both curiosity and horror. As much as they hoped to catch a glimpse of the Beacons, they were just as terrified by the thought that they'd shown up so close to home.

Intrigued by Red Bean's loud knocking, neighbours poked their heads out of their doors. She was wearing a yellow jacket on top of her red skirt. The snow hadn't stopped, and she looked more like a woman in a painting than a real person. There was a man with a camera standing to the side of her. She must've come here straight after her programme.

Dong Ji was the one to open the door.

'I'd like to interview you about Hua Ge,' Red Bean said. 'The truck driver who turned into a Beacon.'

Upon hearing that, I stepped outside. Some neighbours had ventured out of their houses and were standing a safe distance behind the cameraman.

'How did you hear about this?' I asked Red Bean.

'It is my job,' she said with an indifferent expression. She didn't seem to recognise me from the day we sat at the bus stop together.

'I told the police that I didn't want this to be public information,' I said. 'Please leave before people gather.'

Dong Ji shot me a glare before reverting back to Red Bean. 'If the police told you, then shouldn't you know everything already? We don't want anything more to do with this.'

'I'd like to use your name,' Red Bean said to me. 'I'd

like to put a face to the story. I need your permission for that. I'd also like to hear you talk about your experience. I believe it'll be more powerful this way.'

She was so collected and put together that I began questioning whether it was really the same woman I'd met the other day at the bus stop.

'Please leave,' Dong Ji interjected before I could say anything. 'What makes you people think that you can just come over to someone's home with that camera? Will you be responsible if something happens to us?'

I pulled on Dong Ji's sleeve. She pushed my hand away.

'Shouldn't you be in the Northern District filming those Beacons?' she continued. 'Are you afraid? Is that why you came to us, civilians, who don't want anything to do with this?'

'It is your responsibility,' Red Bean said directly to me. 'You're responsible, as a citizen of Five Poems Lake, to share the information you have with all your people. Look at the sky. It's ten o'clock in the morning and almost completely dark. Don't you think we should all play our part to figure out what is happening to us? To save ourselves?'

'Save ourselves? You really think you're saving us by doing what you do?' Dong Ji said. 'Leave before I call the police.'

Dong Ji pulled me back inside and slammed the door shut.

'I filed the report,' I admitted.

She sat on the sofa and did not look at me.

'I'm sorry I made the decision myself,' I said.

'Are you?'

She stood back up, stormed into the kitchen and started washing dishes at the sink. I followed her.

'I just think it's important to tell the police what I know,' I said. 'Ba would've done the same.'

'Then why didn't he tell anybody about the photograph?'

'I don't know,' I said. 'I don't. But the more we know about the Beacons, the closer we will get to the truth behind the photograph, right? Isn't that what you want? If we keep all the information to ourselves, then we will never be able to learn more about them.'

She spun around to look at me.

'So?' she asked. 'What have you learned? You've learned that the police don't care about us. Did they keep your secret for you? Didn't I tell you that this was going to happen? That those people would show up?'

For a moment, she looked like she was going to go back to washing the dishes, but then she stopped and held her face in her palms.

'Do you have any idea how difficult it was to take care of a child when I was a child myself?' she asked.

I didn't know what to say.

'I have a responsibility to protect you,' she said. 'To

make sure nothing happens to you. But it's so hard when you won't listen.'

'I'm not a child. You don't have to protect me any more.'

'So what am I supposed to do now? Go off by myself and leave you alone? You're my little sister, for heaven's sake.'

'You can start with letting me help with the dishes.'

She turned back to the sink again and picked up a dirty bowl.

'You know,' she said, 'for years after Yeye died, I had to wash the dishes again after you did because you hadn't cleaned off all the soap?'

'You could've just told me to do them again.'

'That's not the point!' she shouted. 'You had to focus your time on homework.'

'It never had to be that hard!' I shouted back. 'I didn't have to finish school. You made me do it, remember?'

She threw the bowl into the sink. It shattered and crashed loudly against the other dishes and splashed water all over her clothes. I stopped talking.

'If I were to do it all over,' she said, 'I wouldn't think twice before sacrificing my dreams for you again.'

Piece by piece, she began fishing out the broken ceramic from inside the sink. Her fingers moved delicately, like she was picking up flower petals.

Then she said, 'But that doesn't mean I can let go of this resentment.'

She lowered her head over the sink and began to cry. Strands of her long black hair trailed into the water.

'I hate myself,' she said, her voice barely audible. 'I hate myself for feeling like this. If I were your mother, maybe I wouldn't feel this way. I don't know. But I'm not your mother. I was just a child too. How I wish you had a mother to love you in the ways I couldn't.'

She wiped her face with her shoulder. It felt unfair, like I was being made to answer for things I never chose.

'You're not the only one who sacrificed something,' I said. 'In fact, you were the one who was given the freedom to go anywhere and do anything but instead chose to spend your days rubbing honey and brown sugar on people who—'

'I'm not the one at fault here!' she shouted. 'I'm not! I've tried . . . I've tried so hard.'

I'd rarely ever harboured any ill feelings towards my mother, but now I blamed her, not for her absence in my life but for the deep pit she'd left in the people I loved, one they felt they needed to fill. I wanted to say something, but my voice got caught in my throat and I felt like I couldn't breathe. It really did hurt to hear about the feelings Dong Ji had kept suppressed throughout the years, but what was more painful was seeing her admit to them. I felt guilty for all the things I couldn't control, but I did not apologise. Remorse felt insufficient and unneeded.

I saw at once how Dong Ji and I had never quite

learned how to be in relation to others. We were clumsy, either living entirely for another person or completely for ourselves. Neither was a way to exist as a human being in society. And the rift between us was only going to grow wider if we continued to treat each other this way.

I threw on a coat and, before leaving the house, I handed her a box of tissues as gently as I could, hoping that she would understand that I didn't blame her for anything. Outside, Red Bean was gone and the neighbours had all retreated back into their homes. I headed towards the police station.

Gao Shuang was standing in the corridor talking with a higher-ranked officer when I saw him. I strode up and got his attention.

'Who told her?' I asked.

They stopped their conversation and the other officer looked at me questioningly.

'Isn't this Dong Yiyao's daughter?' he asked, smiling down at me.

'Who told her?' I asked Gao Shuang again.

'What's wrong?' he asked.

The other officer's expression reverted into a serious one as he patted Gao Shuang on the back.

'I'll see you at the brief,' the officer said before he walked away.

Gao Shuang pulled me into a nearby room that was empty.

'Have you been crying?' he asked. 'What's the matter?'

'Red Bean came this morning,' I said. 'She wanted to interview me about Driver Hua. She even knew his name.'

'Bastards,' he said. 'They always end up knowing everything that happens here.'

'You assured me that the police would keep it a secret,' I said. 'I made it clear that I didn't want to be identified. I made a promise to Driver Hua's wife. I feel terrible.'

An expression rose to his face that was at once innocent and guilt-ridden. I knew that he was not the one who leaked the information, but I couldn't stop myself from unleashing my anger onto him. I was also aware that he had the habit of blaming himself for the failures of those around him. The last thing I wanted was for him to feel guilty: he had already been carrying the heavy load of Ba's death on his shoulders. I shouldn't add to it, I told myself.

'Did you have a fight with Dong Ji?' he asked. 'Was she mad at you for telling us?'

'You guessed it.' I tried to smile.

'Fuck. This is not the time to fight, you two.'

He ruffled my hair like Ba used to. I brushed his arm away. I was still upset. He opened his mouth to say something but then remembered that the door was open. After poking his head out to make sure nobody was going to come in, he closed the door.

'What is it?' I asked.

'I don't have much time,' he said. 'I need to go look for someone.'

'Look for someone?' I asked.

'One of the officers in my team didn't come into work today,' he said. 'I need to find him. There's something I have to ask him.'

'Is it the officer who whistles when he speaks?' I asked.

'Whistles? I don't think so. He doesn't whistle. It's Big Su. You met him that one time, remember?'

I nodded.

'Yesterday night,' Gao Shuang said. 'He'd already clocked out and gone home, but after an hour, he came back. Something was off. He wasn't acting like himself.'

I waited for him to elaborate.

'He told us that he's a Beacon himself,' he said. 'And then he stopped talking.'

'Is he?' I whispered. 'Is he a Beacon?'

'He doesn't have that sun on his head,' he said. 'If that's what you're asking.'

'It's not *on* their heads,' I said. 'It *is* their heads.'

'Well, he looks the same to me, only he stopped behaving normally. He's got this look in his eyes. Vacant, like something inside him is gone.'

'He told you that he's a Beacon?'

'Yeah,' he said. 'He also said that he saw a Beacon when he was young. I think he's lost it.'

'He did? How long ago?'

'He said that it was when the sun first started to disappear. He was still living on the boat. So it must've been something like twelve years ago.'

'Did he say where he saw this Beacon?' I pressed. 'Was it a man or woman? What else did he say?'

'Slow down,' Gao Shuang said, surprised at all the questions I was firing at him. 'Don't think too much about it. He's clearly not himself.'

'Do you know where he might be?' I asked.

'There are some places that I want to check,' he said.

I hesitated briefly before asking, 'If you find him, can you bring him to the pharmacy?'

Gao Shuang shot a glance towards the door to double-check that it was closed.

'Didn't I just tell you that the police are looking for him?' he said quietly. 'Why would you want—'

'That was when Ba died,' I said. 'Twelve years ago.'

'So what?' Gao Shuang said. 'That was when a lot of things happened.'

'The night before he died,' I said, 'Ba told Dong Ji that we could make our own suns. Maybe Ba also saw this same Beacon twelve years ago. Dong Ji is trying to find out more about how he died.'

I continued to tell Gao Shuang about our decision to dig up Ba's ashes and finding the photograph of the Beacon inside. Maybe I told him because of my argument with Dong Ji. Maybe I wanted to do something else behind her back. Plus, I no longer wanted to keep things from Gao Shuang. I was tired of secrets.

He rubbed on his thumb while I spoke, and I could tell that he was trying his best to hide those wounds that never quite healed. I wished he'd let me comfort him, but the only way he knew how to reconcile those emotions was to seal them away.

When I finished, I asked him to tell me everything he knew about Ba's death.

'I don't care if it's the smallest detail,' I said. 'Tell me what he ate that day. Tell me what he was wearing. Everything.'

He plopped himself down onto a chair and stared into nothing for a while. Before long, he began scratching at his neck.

'I don't know what he ate or what he was wearing,' he said. 'But I guess you know about the missing persons cases.'

He looked up at me and I nodded.

'The first person who went missing was a woman. Her husband came in to file that report. Turned out she just ran away with her lover. But then, the husband disappeared too.'

'I know about that. Anything else?'

'After that, a student went missing. The Commissioner pulled the file before we could see it.'

'Why?' I asked.

He leaned forward in his chair and continued to scratch his neck.

'Why?' I repeated.

Gao Shuang pounded the table with his fist. The sudden and loud sound startled me.

'I don't know, OK?' he said. 'I've always wondered whether it was my fault that he died.'

His fingers ruffled vigorously through his hair.

'You didn't kill him,' I said, noticing that I was hurting him just by pressing the matter.

'I just have the feeling that he was doing something alone and I wasn't there to help him.'

'Did you see Ba's body?' I lowered my voice. 'Were there really no signs of struggle?'

'Your grandfather wouldn't let us see it.'

He straightened his neck and said, imitating Yeye's tone of voice, 'Son, there's nothing pretty about it.'

The resemblance made me smile.

'Not bad,' I said. 'You're getting better at this.'

He stood up and told me that he really had to go. The longer he waited, the more likely someone else was going to find Big Su. He couldn't promise that he'd be able to bring Big Su to us, but he agreed to try his best.

'He's not as much of a jerk as he seemed,' Gao Shuang said before he left. 'He is just carrying a lot. We all are.'

When I returned, Dong Ji was asleep in the room that we used to share. I pulled out an old duvet and spread it over the sofa. It was a little past four in the afternoon, so I brewed a large pot of chamomile tea and sat by the window.

As the last of the sun's shafts moved away from the deserted streets, I saw a woman walk past. Her back was hunched and she was hugging a loaf of sandwich bread close to her chest. She walked a little farther before ripping open the plastic packaging and taking out a piece of bread. She scarfed down two slices, barely chewing. It was like she cared about nothing but the bread in front of her, like insatiable hunger was a vacuum inside her. As she pulled out another slice, I saw a light bubble out of her mouth and set the bread on fire. She dropped the charred slice and took another from the pack. The light bled out some more and the same thing happened. Then a man ran up from behind her and swung a shovel into her back.

I watched her figure collapse on the ground, the light still spilling out of her gaping mouth. The rest of the bread rolled into the snow. I couldn't tell whether her eyes were open or closed. The light had obstructed my view. The man threw the shovel to the side and began kicking her torso. She did not seem like she was in pain. I'd like to believe that was why I did not stop him, but if I'm honest, it was because I was afraid. The moon was partially covered and all I could see in the darkness was the shadow of the man, tall and fat, with a long ponytail. I couldn't see his face, but he was easily identifiable, an electrician. Sometimes, when I bumped into his wife at the market, I would greet her, but I did not know their names.

The man stopped abruptly, as though he'd remembered something. He took a step backwards, in a state of disbelief, and took off running towards his house. The woman lay still on the ground, unconscious, I presumed. A few moments later, I saw the man approach again. This time, he prodded the woman with his leg and, having verified that she wasn't going to move, he picked up what was left of the bread and ran off.

Nobody came out of their homes. They must've watched from their windows. The snow continued to fall and, just a short hour later, it buried the woman's light.

8

The rest of the sun disappeared, and Five Poems Lake fell into total disarray.

Even though this day had been inching closer to us for years, its arrival still came as a shock to the town. Perhaps it was because the sun had been fading at such an abnormally fast rate in recent days that we were hardly spared a moment to prepare for the morning that it would ultimately abandon us. Then again, I reckon we could never be ready for something like that, even if we were given an eternity.

To add to the chaos, the Beacons were everywhere now. Like an unstoppable disease, their numbers multiplied exponentially. Even now, there were four Beacons standing in front of my door, like lanterns hanging along the pavement. They were our light source now, driving away the darkness that'd felt so inescapable just a few days ago. Despite the disappearance of the sun, the temperature did not become colder and even went up a little, though there was no certainty as to how long this warmth would last. People went into the police station every hour to report new cases of their loved ones and neighbours turning into Beacons. Every incident happened in the same manner. The light came from within their throats, they all said, up from somewhere deep inside the dark caverns of their bodies. Had I known that things were going to escalate so quickly, that other people seemed to feel no hesitation telling the police, I wouldn't have filed that report myself. Every day, I was afraid that Driver Hua's wife would knock on my door and tell me how angry she was that I broke my promise.

Naturally, people had stopped going to work. I'd closed the pharmacy. The only people still courageous enough to show up at their jobs were a couple of police officers and farmers who had stock of perishable food. As for the other citizens of Five Poems Lake, they had either packed up their lives and run away or hoarded food and locked themselves at home. The few buses going out of town were always packed now, and those who couldn't get on the buses became a constant trail of people trekking towards the desert.

As the Beacons multiplied, the demonstrations began gathering force. It started when information regarding a Beacon who had been secretly detained by the Eastern Police had been leaked by a young officer. The justification was that they wanted to learn more about what they called 'the nature of the Beacons'. The police really had no control over their own secrets.

What a mess, I thought. It wasn't like this when Ba was still alive. It hadn't even been so awful when the sun first started to disappear. Or maybe it'd always been this way, and I was just too young back then.

Here in the Eastern District, we all knew the Beacon that they'd detained. I'd never heard anyone refer to him by his real name. We all called him Mutt. He did not have a stable job, a wife, or any family, and rarely ever left his home. He lived on a sum of money that his parents had left him before they'd died. On the odd occasions he did go outside, he spent his time insulting passers-by just to provoke them into getting into a fight with him. Sometimes, Mutt would get beaten up so badly that the police would have to help him to the hospital. They must've taken him because they thought nobody would've cared. The police could be such idiots sometimes.

Two days ago, we woke up to flyers everywhere – taped to lamp posts, doors, walls. A photograph of Mutt was cut in the middle along his neck and pieced back together again with his glowing head on the bottom and his body standing on top of it. His head was confined inside a cage that somebody must've taken a lot of time to draw. On the legs of his trousers, they had written 'I'm Innocent' and 'Release Me' vertically in big characters. Then, the demonstrations began. Even though it didn't seem like people had much sympathy for the Beacons, I suppose they'd still believed them to retain something human. I wasn't convinced that these demonstrators were really fighting for Mutt. They fought because it was the only thing left to do. They felt wronged – we all did, in one way or another – and they blamed it on the police. But what could the police do? They were just people, too.

Amid all this mess, very few people were talking about the sun any more. We used everything to distract us from what was really happening; the fact that we were going to die from the darkness, the cold, from starvation, from disease, from a lack of nutrients and, for the least fortunate, from the rot in their own minds. We couldn't look death in the eye and ask why it'd come for all of us. That much had been clear for a long time now. We scurried away at the very thought. We directed our anxieties towards the practical rather than the existential, like when our electricity was going to be cut or when our water was going to run dry. Our minds stopped short of thinking about death.

Dong Ji and I confined ourselves at home and spent our time sorting through Ba's old belongings. We had been putting it off for years, leaving the boxes to collect dust. It was difficult to revisit that pain, to think about all these years without Ba. But digging up Ba's ashes must've opened up a pathway, a door towards reconciling with that pain that we'd never properly addressed.

For a man who'd never seen a need for excess, Ba really did leave a considerable collection behind. I suppose it is impossible to exist in this world, even if it was for a day, without shedding some of your skin. Most of his possessions were ordinary – the leather-strapped watch, clothes, a metal flask still half-full with liquor, an entire bag of lighters, among other even less noteworthy objects. There were no journals, notes, or records of any sort. There were photographs of all kinds of things – birds, butterflies, people, objects – but no more Beacons, and nothing else out of the ordinary.

'Do you think they held a big funeral for Miss Pan?' I asked Dong Ji, who was cutting a sealed cardboard box open with a knife. It seemed like so long ago when she came into the shop asking me to cook bird's nests for her, but it'd only been two weeks.

'It was scheduled for yesterday,' she said, 'but with everything that's been happening, I don't know.'

'I hope they went through with it.'

'Me too,' she said.

'What about the baby? Did they hold a funeral for the baby too?'

'I don't know. I don't think so. But maybe. Actually, I'm sure they did. It's the only thing that makes sense.'

'I don't think she had thought of a name for the child. What would they write on the headstone?'

'I don't know.'

Dong Ji peeled the rest of the duct tape off the box and opened the flaps. She lifted some sandals out and then flipped the box over. Pens, keys, screwdrivers and a hammer fell onto the floor. In the distance, we could hear protestors shouting Mutt's name in front of the police station. It made me feel sorry for Mutt. They knew nothing about him. None of us did.

'In a way,' Dong Ji said, 'it's a good thing Miss Pan never lived to see the world like this.'

'I keep imagining that she'd smile about all of this. It always felt like nothing could ever bring her spirits down.'

All of a sudden, my chest tightened, but I took a deep breath and drank some water instead.

'Oh!' Dong Ji said. 'I have something to lift our spirits.'

She stood up and grabbed her purse from the bedroom. She stretched her hand into the outside pocket and found a handful of caramels wrapped in golden foil. She gave one to me and splayed the rest out on the floor between us.

The sweet and creamy caramel melted in my mouth and stuck to my teeth. I immediately went for another one. Dong Ji started giggling.

'What?' I asked.

'I was just thinking about the summer when we climbed onto the roof every day and melted caramel onto the tiles. There were all these squirrels up there eating the caramel. And then the ants came. Yeye was so mad.'

'You had to clean all of it because you were the older one.'

'Burnt caramel and squirrel shit. It all started to look the same to me.'

We both laughed, exposing our caramel-covered teeth.

'You know,' Dong Ji said, 'you have Ba's feet.'

'I do?'

We both looked down at my bare feet.

'I don't remember what his looked like,' I said, feeling self-conscious.

She reached over and squeezed my foot.

'Go put on some socks,' she said. 'They're cold.'

'I like it this way.'

'It's bad for your health.'

She stretched her arm towards the sofa, pulled the blanket off and wrapped it around my feet. The lemon-coloured fleece was soft and warm against my skin.

We heard a loud thump. A Beacon had tripped over and banged himself against our window. We pushed ourselves up immediately and climbed onto the sofa to look outside. He was kneeling on the ground, his body leaning limply against the brick wall. His little sun was about the size of a basketball. I could feel the searing heat radiating through the glass.

Apart from when Driver Hua turned into a Beacon, which had happened so quickly and I'd run away before I could get a good look at him, this was the closest I'd ever been to one of them. The light was so blinding that I had to squint and block it with my palm. Through the gap between my fingers, I observed him. He was thin with wide shoulders. All he wore was a T-shirt and some dirty black jeans. I couldn't be sure, of course, but the little sun looked like it was too heavy for his body to support, making the movements of his limbs ungainly. His legs continuously struggled to push him upright, but the weight of the light kept him on the ground.

I opened the window a little.

'Hello?' I said in a whisper.

'What are you doing?' Dong Ji slapped my arm and pulled the window shut. 'Don't provoke him.'

'We've got nowhere with the photograph.' I pointed towards the altar. 'We've found out nothing. Let me talk to him.'

'Here,' she said. 'Let's trade spots. Let me try.'

'I can do it.'

Reluctantly, she took my hand off the window. I turned around and picked up Ba's hammer from the ground before opening the window again.

'Do you need help?' I asked the Beacon.

As expected, it was as though he couldn't hear me at all. Dong Ji found a long umbrella leaning by the door and stuck it out the window. She poked him on the shoulder. For a moment, I saw his sun float in the air above his shoulders and, in an instant, it was like it'd become light as a balloon. His body was pulled up by the little sun as it rose, until soon he was standing upright again. The little sun moved and his body followed. It was a subtle thing, I felt, like he had no power over himself.

The Beacon was now standing right in front of our window like a lamp post, his body facing towards us. The light from his sun brought out the colours of everything around him.

'We'll go somewhere far away from all of this,' Dong Ji said to me.

'We'll die out there,' I said.

'I'd rather die trying to live than live trying to die.'

I couldn't believe that she'd come up with something so wise and irrefutable. Had she been preparing for this moment?

'But this is our home,' I said.

'We can find a new home. As long as we're with each other, we can make a home in the desert if we want to.'

'The sun is gone, Dong Ji. It's gone everywhere.'

'You don't know that!'

'You must feel it,' I said. 'Don't you? I feel cold inside out.'

'Look at what's happening to this place,' she said, pointing at the Beacon outside the window. 'Can you live with them?'

'But what about the pharmacy?' I asked.

'Look at him!' she shouted. 'Do you think he needs medicine? He doesn't even have a mouth. Don't let fear beat you down now. We'll have each other. Remember when you were little? You'd always tell me that you wanted to go out there with me.'

'But maybe the Beacons will save us,' I said. 'There's light and there's heat and—'

'What if we turn into Beacons?'

I was quiet for a moment.

'You won't regret it,' she went on, like a store clerk who'd just made a sale. 'When I moved into the dorm, I thought I'd miss home so much that I wouldn't last a day, but look, I've managed, haven't I?'

'You know this is not the same,' I said.

All my life, I'd deferred to Dong Ji. Apart from filing that police report about Driver Hua, I couldn't think of the last time I'd openly gone against her will.

Dong Ji spent the rest of the day packing for us. She calculated the weight we were able to carry and put together two backpacks of things that she deemed essential to our survival. I counted the days in my head. No matter how meticulously she planned out our provisions, they wouldn't last us two weeks. She must've known this too, but neither of us said anything.

The protests outside the police station grew progressively more heated as the hours went on. The crowds stretched all the way to our shop, and more people were joining from all the other districts. I worried for Gao Shuang. We phoned his desk. Nobody answered. Dong Ji considered going to look for him at the station, but with the hordes outside, they wouldn't have let her in. I prayed that Gao Shuang was safe and that he could escape before they stormed into the building. I wondered how far he'd got in his search for Big Su. Dong Ji was becoming restless, so I tried to assure her that Gao Shuang was a trained police officer, but my own words came out shaky and unconvincing.

The camellia was reaching its limits at last. Digging it out of its pot must've added to the trauma. Camellias were tough plants, but sheer resilience wasn't going to save it any more. The last leaf, which had been hanging on for much longer than the others, finally fell.

It was upsetting, seeing the camellia die. It broke me. While I was sweeping the leaves, an immense, awful feeling overcame me and took control of my body. It paralysed me and I sunk to the floor. For some time, all I could feel was fear. The fear came from somewhere inside me where there was no logic, no language, no order. Never in my life had I been so afraid. I sat there for God knows how long until Dong Ji came out to make lunch and found me, teeth clenched and stiff like a rock. She covered my back with the same lemon-coloured blanket and guided me to the sofa.

'The leaves fell,' I said. 'They're not growing back. It's dead.'

'Here,' she said, handing me a cup. 'Take a deep breath. Drink some water.'

'Did we kill the tree? We killed it, didn't we? Even the camellia tree is dead.'

Dong Ji held me in her arms and ran her fingers through my hair.

'Oh look,' she said. Her voice held a watery texture. 'It's all messy.'

The next thing I knew, she was sitting behind me, one leg hanging on either side, her thin fingers holding a comb and running it through my hair. Her movements steadily lulled me. I focused on the sensation of the little teeth massaging against my scalp. She combed my hair for a long time, until I was no longer shaking. Tiredness fogged over me. The soft pressure from Dong Ji's body and the warmth of the blanket cocooned me in a comforting embrace. I wasn't sure when I fell asleep.

I woke to noises of things breaking and with a dreadful pain in my temples. The protests outside had grown violent. They were throwing things at doors, windows, buses, police cars. I asked Dong Ji what had happened while I was asleep. She told me that the police had managed to evacuate the building. The police's decision to run away fuelled more anger among the protestors. Even I felt disappointed in them; Ba would never have deserted his post, I was sure. Even so, a small part of me felt relieved that Gao Shuang was safe somewhere else.

While I was asleep, Dong Ji had dragged the camellia

plant outside so that I didn't have to see it any more. She'd barricaded the back door with our dining table and chairs. With all the commotion she must've made, I couldn't believe that I hadn't woken. The stress that'd been building up throughout the past days must've really caught up with me. Even after the nap, my muscles still felt weak and my mind was in a haze. Now that day and night were no longer consequential, sleep had lost its meaning too.

 I helped Dong Ji push the sofa against the door connecting our living spaces to the shopfloor. She returned to the kitchen and started washing some spoons. An enormous pot sat atop the stove.

 'What are you cooking?' I asked.

 'Ginseng congee,' she said without turning her head. 'I used the rest of the ginseng we had left in the shop. It should give us some energy. Are you hungry?'

 'A little. Let me know when it's ready.'

 I dragged myself back to the living-room window and brushed aside the curtain. I saw the back alley packed with protestors carrying all sorts of weapons. The garbage bins and bicycles were all destroyed. Some of the windows, too. Among them were a dozen or so Beacons. They seemed to pay no heed to the people around them. Most of the protestors avoided veering too close, for fear of burning themselves. Towards the end of the alley, I saw a well-built man with a Beacon – a child – sitting on his shoulder. The man was yelling something furiously with his fists pumping in the air while the little Beacon sat atop his shoulders. Then, I saw the man trip and fall forward with the Beacon, out of sight. I kept my eyes on the empty space where they'd fallen.

'Here,' Dong Ji said as she handed me a bowl of congee and a spoon. I'd been so focused that she had to tap me on the back of my head to get my attention.

Just as I was taking the bowl from her, we heard the sound of glass shattering. It came from the direction of the shop. We froze. I heard a woman's voice.

'Medicine!' the woman shouted. 'They're hoarding all this medicine even though Five Poems Lake is dying!'

Dong Ji stood there staring at the door that led to the shop. I picked up Ba's hammer from the corner of the room and stuffed it in her hand, switching out the bowl. When we were children, I'd often been told that I was the braver one out of the two of us. Dong Ji was frightened of many things – the dark, large dogs, swimming, thunder, riding a bike down a hill, confessing to the boys she liked. Unlike everybody else, I never thought that she lacked courage. I looked up to her for being afraid of those things. She admitted to her fears, which made her much braver than me, who avoided talking about mine entirely. It was regrettable that, as we grew up, I'd noticed she'd become more like me.

We quietly made our way towards the sofa and hid behind it.

'The gloves!' I remembered. 'Miss Pan's gloves. They're out there.'

'Not so loud!' Dong Ji whispered, her voice shaking so much that I had to read her lips to understand what she was saying. 'Leave them.'

'But what about all our medicine?'

Dong Ji deflected my question with a series of frantic shakes of her head.

'I think there's only one woman in there right now,' I said. 'If we don't stop her, she's going to take everything. We need emergency medication.'

She grabbed my arm.

'Give me the hammer,' I said.

For a moment, she looked like a little girl. I'd never seen her like that.

'I'll come back immediately if anyone else shows up,' I reassured her.

I took the canvas bag that hung on the coat rack. I turned it upside down and emptied its contents. We pulled the sofa back a little and I wedged myself between it and the door. I took a deep breath and turned the knob as quietly as I could. On the other side, I could hear more glass and wood being smashed. I cracked the door open and crouched as low as I could so that when I entered the shop I would be hidden behind the counter, out of sight. The quilted curtain I'd hung on the door frame was so heavy that I had to put the hammer down to push it aside. As fast as I could, I slid into the shop and sat with my back against the counter, making sure I remembered to pick up the hammer. The store was rather dark; the light from the Beacons had not reached inside. My heart felt like a swollen balloon, blocking my windpipe. The woman was talking to herself.

'What the hell is this?' I heard her say. Her breathing was heavy, like she was afraid too. 'Everything looks the same. No matter,' she said. 'I'll find a use for it.'

Her voice came from the other side of the shop, so I crawled to the end of the counter to get the bandages and iodine. I retrieved them easily enough, but the problem now was that I couldn't get to the fever and digestive medicines without crossing the room. I couldn't believe that I didn't have an emergency pack ready at all times. I ran a pharmacy, after all. I felt stupid, inexperienced. I decided that I'd make do with some goldthreads, catnips and perilla leaves, but even those ingredients couldn't be accessed without exposing myself somehow.

I couldn't afford to wait any longer. At this rate, she was going to comb through all the cabinets. She would take all the medication I needed before I could get to them. A ray of light shone into the shop and moved away after a few seconds, like a car driving by at night. It must've been a Beacon walking past the entrance. I looked around me. On the shelves above, there were glass jars full of dried wood ear mushrooms, goji berries, caterpillar fungus and jujubes. I reached for the jar of jujubes and quietly poured out the dry and wrinkled fruits. They scattered like large red beads onto the dark floors. Hammer in one hand and the empty glass jar in the other, I turned around and stretched my head above the counter. The woman had her back towards me. It was dark, but I could see the outline of her body. I was relieved that she was rather small, about the same height as the Su girl. She was rummaging through the drawers of the apothecary cabinet and stuffing ingredients hastily into a bag, mixing all of it together. I stood up and advanced towards her as soundlessly as I could. The store was in ruins. The cabinets were all broken. The shattered glass on the floor looked like spiderwebs.

I managed to walk all the way up to her without being noticed; any noises I made were masked by the madness outside. I was soon close enough to see the folds in the woman's collar and the tips of her ears, even in the dark. Her hair was cut short. There was a tattoo of a needle on the back of her neck. The tip of the needle was pointing towards the bottom of her scalp.

Even though I was right there, close enough for her to feel my breath if I hadn't been holding it in, I realised that I had no idea what to do. My arms went heavy, like my bones had slipped out of me. I couldn't swing the hammer. The glass jar in my hand was useless as a flower and I let it fall and shatter on the tiled floor. The woman spun around and I saw that she had a butcher's knife under her armpit. She pulled it out and pointed it at me.

'Stay back!' she shouted. 'This is a sharp knife! I just sharpened it!'

Out of nowhere, maybe because my body sensed that my life was in danger, I was able to move my arms again. I lifted the hammer up to communicate that I, too, had a weapon. From our clumsy movements, it was clear that neither of us knew how to use the tools in our hands to hurt others. She held her bag in front of her chest like a shield and took a hesitant step forward, crushing some broken glass with the soles of her boots. From the corner of my eye, I saw Dong Ji appear from behind the quilted curtain.

'Go back inside!' I shouted.

Another Beacon walked by the store and illuminated Dong Ji's face. I saw her eyes ricochet between me and the woman like bullets. The woman swivelled her body and pointed the knife at Dong Ji.

'Stay there!' the woman said, her shaking voice betraying her words. 'Don't test me. I could do anything!'

'Just take the medicine you have and leave us,' Dong Ji said.

'What's that in your bag?' she asked me.

I tossed it over. Some bandages and iodine that weren't worth protecting with our lives. Just as she picked up the bag, the streets grew bright and hot as though the sun had returned in all its glory. The three of us looked outside. People, hundreds of them, everyone I could see, started to turn into Beacons, like streetlights being turned on. Their voices became muted as their mouths faded into their suns.

The woman with the knife screamed at the sight and ran out of the store. I saw her bump into a Beacon as she fled. She dropped her bag, but she did not stop.

I was hypnotised by the beauty of this transformation, and irresistibly drawn to its warmth. It was like discovering fire for the first time. Even so, what I did was run away and hide, because somewhere inside me, in a place where all that existed was raw, animal meat, I was acutely aware of its ability to destroy.

I grabbed Dong Ji's arm and together we ran back into the living room. We curled up behind the sofa, making ourselves as small as we could. We held our breath and waited for something to happen, praying that nothing would. I don't know how much time passed before the last sounds dissipated and the outside world became silent. The light from the Beacons seeped through the curtains and illuminated parts of the room. My eyes came to rest on the bowls sitting atop the dining table. The border of the window sliced the white bowls diagonally down the middle, leaving half of them in shadow. Briefly, it reminded me of the past.

For what felt like a long time, neither of us did a

thing. My mind travelled to different places, but eventually it came back to the bowls. In a way, I thought, our bodies are just containers. I'd always assumed that we have the capacity to absorb the world, to endlessly store everything that we come across. But as we sat there in silence, I began to question whether there were, in fact, things that could overflow from us. Is that what the Beacons were: people who had become too full?

I don't know whether I was too afraid, too exhausted, or too confused, but I couldn't bring myself to move. I wanted to sleep again, to let time pass without me knowing. I could've stayed that way for ever, sheltered by this house I knew so well, huddled with Dong Ji on these floors I'd swept every day, basking my feet in the light of the Beacons. I really could have stayed that way, but Dong Ji broke the silence and, in a slow and hushed voice, asked me whether the sun had been gold or silver.

'I don't remember,' she said. 'All those years ago, before it began to fade, what did it look like?'

'I think it was white,' I said.

'That means it must've been more silver.'

'I guess so.'

'Odd thing,' she said. 'I'd always thought it was gold.'

I had a dream. I was standing at the very top of a giant, triangular structure made out of sand. Dong Ji was running towards me, yelling something I couldn't hear, and

I called back to her, telling her to hurry up and join me. The size of the structure kept growing and Dong Ji was unable to keep running, though she was trying with all her might to reach me. I watched as she became progressively smaller, until she turned into an ant. Alone at the top, I couldn't stop crying, but I did not make an effort to descend. I stayed there until day somehow became night, and the tiny black dot that was Dong Ji had vanished into the darkness. All of a sudden, I noticed that the darkness that surrounded me was not the sky. I was underground. I was an ant too. I wiped my tears and began desperately searching for Dong Ji, but I couldn't differentiate between all the swarms of black ants around me. They formed a large, pulsating mass, and I could hear their legs – my legs – tapping against each other's bodies like teeth chattering.

Dong Ji woke me up to tell me that the camellia plant was burning. A Beacon had sat down next to it, setting it on fire with his little sun. He hadn't done it on purpose, she said, she didn't think he'd meant to hurt it. She didn't think he'd meant to do anything. But the fact was that the tree was burning down. By the time I made it to the living-room window, the fire had already eased into a weak flame, flickering at the tip of one of the branches. There were two Beacons sitting there. From their clothes, I could tell that they were a man and a woman. She leaned her little sun against his shoulder. The heat must've been burning away his clothes and scorching his skin, but he didn't seem to feel any pain. Their suns merged together into one single light, the shape of a peanut.

I opened the window a crack and I could hear a few voices yelling something in the distance – far quieter than before. The air smelt like smoke. I stretched my head out and looked down the back alley. The police building had been set ablaze.

'Dong Ji!' I said, before realising that she was already leaving from the back door.

I sprinted after her, as quickly as my legs could take me, as she ran towards the station. I had to stop her before she stormed into the burning building. We both knew that the police had evacuated it, but we ran as fast as we could anyway. Almost everything that defined our father's life was in that building. I didn't want to see him dying all over again.

Dong Ji ignored my calls from behind asking her to wait for me. Giant clouds of stone-coloured smoke rose from the building, pieces of the walls chipped off like little meteors, and I could see bright orange flames fluttering inside the windows. It seemed that the back gate of the station was locked. Even the security guard had abandoned his duties.

I caught up with Dong Ji and we circled around to the front. A fire truck pulled up but only two firefighters came out. The rest of them must've turned into Beacons, or abandoned their jobs and hidden in their homes. Even so, we were relieved to see them. Dong Ji urged them to hurry and asked whether they could save the building from burning to the ground. Both men ignored her, focusing instead on unwrapping their hoses. The red paint on the truck had faded to the colour of a ripe peach.

I grasped onto Dong Ji's arm. We both knew there was nothing we could do. We were not Ba. We didn't have solutions to everything. Sometimes, all we could do was stand and wait.

The Beacons were entirely unperturbed and were uniform in their silence. They wandered the streets in a leisurely manner, like wild dandelions swaying in the wind.

The firefighters had not gone into the building. They were both manning the hose on the truck, spraying a powerful stream of water at the flames. The emblem on their uniforms caught my eye; they had been dispatched from the Northern District.

'It's not working,' Dong Ji said to the firemen. 'Don't you have more hoses?'

They ignored her again. Their expressions were severe, hard yet brittle, as though they could withstand the heat and the cold but with a push they might shatter. Their held their gazes at the burning structure and didn't turn to look at us.

The station was entirely shrouded in smoke. It had become unrecognisable, beyond saving. I thought about Ba's old desk and chair in flames and all the certificates and medals inside the cabinets along the hallways, all burned to ash. It made me angry. I wanted to scream at whoever had done this. They didn't have the right to destroy something so precious to us. Maybe if my skin was just a little thinner, a sun would burst out of me too.

The wind picked up and the fire erupted into a roar. The smoke darkened to the colour of charcoal. The wind blew towards the lake, carrying the fire and smoke with it, and all that could be heard were the flames.

It was then that some police officers came back from wherever they'd been hiding. They must've wanted to save the building too. Or maybe they felt safe now that most of the protestors had turned into Beacons. Whatever it was, they were too late.

'Have you seen Gao Shuang?' Dong Ji ran up and asked one of the higher-ranked officers.

He blinked at us, as though unable to comprehend what reason we had for talking to him.

'No one is here,' he said. 'Go home. It's not safe out here.'

He turned away and looked around. On the main road, it seemed like there were just a handful of protestors left who hadn't turned into Beacons. One of them was taking off his shirt and throwing it into a pile of clothes that was also on fire. The rest of them were smashing things and yelling at the officers, venting their anger like bursting balloons. The police had already started to round them up. A smaller group of officers directed the Beacons towards the lake, making them put their hands behind their backs, threatening them with their batons.

The higher-ranked officer shouted at his subordinates.

'My team, stay behind and help,' he ordered. 'The rest of you, continue your search for Big Su!'

Dong Ji pulled on his sleeve. He gave an impatient grunt, but then looked at us with a more sympathetic expression.

'We're all very upset,' he said. 'Not to mention overwhelmed and exhausted. I can only ask you to keep yourselves safe and go home. I don't know where Gao Shuang is. I'm looking for him too, and I must urge you to let us know immediately if he shows up at yours.'

The officer didn't wait for us to say anything else before marching away to his team. Why was he looking for Gao Shuang? Why wasn't Gao Shuang with them? Was he still looking for Big Su by himself?

The two firefighters persisted, sweat dripping down their necks. We watched them fight the fire, paying no heed to the officers' voices pressing us to go home. They kept the hose pointed at the flames until the water stopped.

'What happened?' Dong Ji asked.

'The tank must be out,' I said.

As I turned my attention away from the police station, unexpectedly, I found myself less afraid of the Beacons. I was even drawn to them. Before, I'd failed to recognise anything in them, as though with the release of their lights, they'd become something entirely different from me. Now that had changed: I felt a strange closeness to them. In some capacity, I could even be made to believe that I had inside me just as much heat as they did.

I'd always known that if the sun was going to disappear, there was nothing we could do about it. We had no control over what nature had in store for us. Now, looking around me at all the Beacons, I realised that we had no power even over ourselves.

Our consciousness must've had a life of its own, and like the world around us, it was burning.

Later that day, Gao Shuang turned up at the pharmacy. By then, there were too many Beacons bubbling around Five Poems Lake to count. Their lights smeared across our town like spilt water, seeping even into the most remote of the farms. The ice on the lake had begun to melt. The air was hotter than it had ever been. Night had become a distant concept, something from a storybook that we could only imagine. There weren't many dark places left in the entire town; the Beacons had been drawn to those last few shadowy crevices. I'd just seen a dozen of them cramped under an awning when Gao Shuang arrived.

All the light made it so that we couldn't be next to a window without having to squint. We'd spent the past decade preparing for darkness, readying ourselves for the day that the sun would be gone. We had trained ourselves to clutch onto any vestige of warmth, any hint of light, any semblance of hope that darkness would be a little delayed. We'd prayed for this day to come, when there would be light again and the lake would return to its gentle flow. But I couldn't even open my eyes all the way to see our prayers answered.

Was this our own doing? Did all that hunger for light manifest into light itself? If that was the case, no matter how I thought about it, the sacrifice seemed too great. The Beacons were not the sun. The sun wasn't something that could be vomited out of our mouths.

Might Dong Ji or I turn into Beacons ourselves? As far as we knew, there were no signs, nothing that could prepare us for it. I spent half a day in my room etching out a letter for Dong Ji, in preparation for the worst.

Dear sister,

When I was eight, Yeye gave me my first lesson on medicine. Now that I think about it, the lesson had little to do with medicine. He showed me how to use a scale and lectured me on the importance of accurately weighing out each ingredient on the prescription. If we inadvertently measured too much of something, he said, the consequences could be dire. Patients have lost their lives because of the negligence of pharmacists. It took me a whole week to learn how to use that old scale of his. By the time I finally got the hang of it, I'd forgotten why I was learning how to use it in the first place. Recently, I've remembered again. I think what he really wanted to teach me was the responsibility we held for other people's lives.

What a burden, really. I'd like to think that we are only responsible for our own lives, that we alone are to blame for what happens to us. I would like us to live bravely, independently, for ourselves. But that would be selfish, wouldn't it? Still, when I think about what I want for you, I won't hesitate to say that I want you to live for yourself and forget about everything else you think you need to care for. So what is preventing us from choosing the easy way out? Is it love?

I've recently been thinking a lot about us. Have you realised that, in a way, we've traded lives? Like you gave me some of yours and I gave you some of mine. How stupid is that?

Don't be sad for me. If the end of my life means that yours might continue a little while longer, I'll happily sacrifice myself for you, just as you would do the same *for me. Knowing that you are warm gives me light. It looks like Yeye's lesson has stayed with me, after all.*

*With love,
Your little sister*

I folded the letter into a square and stored it in the bedside drawer.

Everything in the pharmacy was broken in some way. Even the cabinets that'd seemed all right at first glance turned out to be chipped or scratched. To tell the truth, I didn't know where to begin to put it back in order. I started with sweeping the glass on the floor, but every time I thought I'd got it all, I would feel another piece crunch under the sole of my shoe.

I'd been staring at the dried herbs scattered on the floor, feeling frustrated at my meagre progress, when I saw Gao Shuang charge in through the door.

'Where were you?' I asked. The concern that had built up inside me flared in my harsh tone. 'Where the hell were you?' I repeated.

'Are you OK?' Gao Shuang asked. Beads of sweat dripped down his chin. 'Is Dong Ji here with you?'

'I'm fine. We're fine.'

'I need your help,' he said as he turned to look down the street. 'Wait here.'

'What on earth is going on?' I asked, but he had already run off again.

'Was that Gao Shuang?' Dong Ji asked from behind the quilted curtain. She pushed it aside and burst into the pharmacy. 'Where is he?'

'He went outside again,' I said as I squeezed my body out from behind the counter. 'I was just about to follow—'

Dong Ji hurried out the door before I could finish my sentence. The Beacons on the streets were facing different directions like passengers in a crowded bus. One of them turned her body towards me. I flinched.

In a few seconds, Dong Ji was back with Gao Shuang. Big Su and his sister were following behind them. The Su girl must've also had trouble sleeping. Her eyes were swollen and tired. The skin around them had taken on a brownish tone.

'What are you doing?' Dong Ji asked Gao Shuang at the door. 'Why are you with these people? You're not bringing anyone inside until you explain to me why.'

'What the hell happened here?' Gao Shuang asked, as though he'd only now noticed the state of the pharmacy.

'Somebody came and took everything,' I said and gestured towards the mess around us – the shattered glass, the ransacked cabinet, the spilt herbs. It made me want to cry.

I pointed to Big Su. 'Where did you find him?'

'Can we talk about this in the living room?' Gao Shuang whispered. 'The police are still looking for him.'

'Has he mentioned the Beacon again?' I asked.

'You know what's going on?' Dong Ji turned to me.

'Do you have any food?' the Su girl asked. She had her thin arm wrapped around her brother's torso. 'My brother won't eat. He hasn't eaten anything. Not even the bird's nests. I told him how expensive and rare it is, but he wouldn't even eat that.'

'What are you two hiding from me?' Dong Ji asked, ignoring the Su girl. 'Don't we already have enough to deal with?'

Big Su had yet to do or say anything. It made me question whether he really needed any help.

'He won't speak,' Gao Shuang said, reading my mind. 'He still hasn't said a word.'

'That doesn't concern me,' Dong Ji said. 'I hardly know him. Frankly, I don't care about him.'

'He might know something about the photograph,' I said to Dong Ji.

Dong Ji gave me a look, imploring me to explain further, but I turned around and headed towards the living room. Gao Shuang lowered his head and followed me, leading Big Su across the pharmacy. The Su girl hobbled behind her brother. Dong Ji did not stop them.

In the living room, Gao Shuang asked me if he could try giving Big Su a plate of whatever food we had. I scooped some lukewarm congee into a bowl and handed it to Gao Shuang alongside a salted duck egg.

'What does he know about the photograph?' Dong Ji asked.

'He said he's seen a Beacon before,' I said. I looked at Big Su for him to confirm. 'Twelve years ago, right?'

Dong Ji stood in front of Big Su and asked whether it was true.

'Was the Beacon a man or a woman?' she asked, over and over again. 'What did it look like?'

Seeing that Big Su had no intention to respond, I was going to try asking something as well, but I noticed the Su girl gazing at her brother in shock, mumbling something. I tugged on Dong Ji's hand to stop her questions.

Finally, we all heard what the Su girl was saying: 'You saw her too?'

'Her?' I asked. Though his face was obscured by the light, I was certain that the Beacon in Ba's photograph was a man.

'I thought you were asleep,' the Su girl said to her brother. 'I was sure that I was alone.'

Then, she looked over at us. I hadn't seen it before, but now I sensed a resentment in her voice and a cold look in her eyes.

FIVE POEMS LAKE, BEFORE DONG YIYAO'S DEATH

9

Dong Yiyao jolted awake to a pounding sound on his window. The morning sun smeared a brilliant red across the empty sky. Amu was standing next to the car, pressing her forehead to the cold glass. She seemed beat. Looking at where the sun was, Dong Yiyao knew that it couldn't have been too long since Amu had walked into the desert.

'Open the door,' she said, excitedly. 'I need to ask you something.'

Dong Yiyao unlocked the car doors, and Amu stretched into the back seat and grabbed a can of orange soda. She slugged all of it down in just a few seconds. He got out of the car.

Upon finishing the soda, Amu exclaimed, 'I get it! I figured it out! This is it, officer!'

She threw the soda can on the ground and grabbed both of Dong Yiyao's arms.

'Tell me how your friend did it,' she said, her cheeks aglow under the warm shafts of the sun.

'My friend?'

'Don't you see?' she said. 'I was walking in the sand and it just hit me. Your friend had his own sun! It came from within him. How did he do it?'

Dong Yiyao realised that she was talking about Ma Gang and he was at a loss as to what to say. He still did not understand what she was getting at. If only he knew the answer to her question, then all of this would've been much easier to understand.

Amu closed her eyes and spread her arms open to the sun, releasing Dong Yiyao from her grasp. She took a deep breath before twirling her body back to face him.

'Can you imagine?' asked Amu. 'If you put your friend's sun next to a plant, I bet the plant will grow. If you shine it on an ice cube, the ice will melt. He could've lived for ever, in everything his light touched. I want to be like that.'

'You want to be like that?'

Dong Yiyao couldn't believe what he was hearing. How could Amu have learned nothing from his story? Why would she want to be like Ma Gang? Why did this girl, who was so desperate to live, grasp onto every opportunity to die?

'He made his own sun,' Amu said again. 'That's the answer I've been looking for. I want to be like that. Please.'

'Listen to yourself,' Dong Yiyao said. 'You're taking this too far. I've had enough.'

'Tell me,' Amu pleaded. 'How can I do it? How can I make my own sun? What do I have to do—'

Right as she was saying this, as though God himself had been listening, a light began to radiate from Amu's mouth, just as it did with Ma Gang. Dong Yiyao immediately recognised that it was happening all over again, like a recurring nightmare, right in front of him. He pressed his hands over Amu's mouth, but the heat burned his palms and bled through his fingers.

Stop it, Dong Yiyao prayed, though he did not believe in gods. *Stop it now.*

Amu did not fight the light. Nor did she embrace it. She let it be, as though it really was the sun, a part of the universe, the existence of which did not need to be challenged. She stood there, arms limp against the sides of her body, as her forehead, eyes and her little nose became bleached by the light. Her body was as straight as a column. The pain she'd felt all her life had seemingly dissipated. It was as though something that had belonged to Amu was no longer there. Dong Yiyao hesitated to call it her soul. He felt hot and cold at the same time, like he had come down with an unyielding fever.

He did not know why this was happening to him. He could not fathom what he'd done wrong to allow the same tragedy to play out all over again. Had he not meddled with Amu's affairs, he thought, she would've still been herself. It was his fault, he decided, all too hastily. It was the only answer he could come up with, and he was in desperate need for an answer, even if it meant it was going to break him.

Amu turned around and began walking, the backpack still strapped to her, towards the desert.

Dong Yiyao's body slumped onto the hood of the car as he felt tears run down his face. It was the first time he'd cried since his mother died. Even when Dong Ji asked him, in her silvery voice, when her mother was going to be back from work, Dong Yiyao had held it together, shoved the stinging feeling down into his guts. But what had happened to Ma Gang and Amu broke him somewhere that couldn't be mended. How could he keep his daughters safe now? He felt as vulnerable as a leaf off its branch, a worm out of soil, a clam without a shell.

When Dong Yiyao regained some of his composure, he shook off his shameful moment of weakness and ran after Amu. He caught up to her and picked her up. There were too many reasons why he couldn't let her go away like that. Amu did not fight back. Her light scorched the bare skin on Dong Yiyao's arms but he persisted through the agony, carrying her all the way to his car. He opened the trunk, set her inside, and slammed it shut without thinking twice.

There was nothing he could do to change the past, he told himself. He needed to think rationally about what had happened. If Amu hadn't seen the photograph, would her fate have been different? He was so careless, stupid. Was the light an infectious disease? If that was the case, he knew that he had the only chance to eradicate it.

But what if it was something inexplicable, like the sliver of the sun that'd disappeared? Things around Five Poems Lake were already bad enough because of that. If people were to find out about Ma Gang and Amu, what would happen to this place?

He thought again about his daughters. His responsibility, now, was to protect his family. It was then that he felt his mind clear, and the shapes of his thoughts become sharper than ever. He made a decision. A cruel one. But he had no choice if he wanted to save Five Poems Lake, save his daughters. He had to protect the peace at Five Poems Lake, for their sake. He was the only one who could do it, he told himself. Nobody else knew about Ma Gang and Amu. Nobody could find out about what he had seen. He had to put an end to all of this.

Dong Yiyao drove along the dirt road through the outskirts of town, passing all the farmlands and villages. He drove through the day, taking longer routes, waiting until the last vestiges of sunlight gave way to the night. Amu remained silent and still in the trunk.

Once it was dark, Dong Yiyao drove to the edge of the Eastern District and parked in a field that was overgrown with wildflowers. Under the starlit sky, the little white flowers looked like patches of white mould in the grass. He got out of the car and opened the trunk.

He lifted Amu out, cuffed her hands and tied her feet together with a T-shirt, and carefully carried her to the back seat. He folded the photograph of Ma Gang into two and slid it into his pocket. The corner pressed into his skin through the fabric.

He had to go home first. How many days had it been since he'd slept in his own bed? He wanted to see his daughters and listen to the mechanical sounds of the ceiling fan until the sun came up again. He decided that he'd give himself that. He'd allow himself the brief comfort of bathing in the stillness of the night.

In the shadows of the field, the car shone bright like the North Star.

10

Dong Yiyao had not expected Dong Ji to be awake. The lights were off in the living room and she was sitting on the side of the dining table that was illuminated by the streetlights outside, scribbling on a piece of paper.

'It's late,' Dong Ji said, making no effort to turn around and look at her father.

'So you know it's late,' Dong Yiyao said.

'I'm old enough to decide my own bedtime.'

'I didn't say anything. What are you writing?'

Dong Yiyao put his hand in his pocket and walked up to his daughter. He clasped the photograph between his fingers and, with his other hand, patted the base of Dong Ji's neck. He saw that she was writing a list of some sort, but she turned the paper over before he could read it.

'It's just something I like to do when I can't sleep,' Dong Ji said.

Dong Yiyao sat on the chair next to her.

'Does that happen often?' he asked.

'Not really,' she said. 'Usually it's because Yeye's snoring is so loud that I can hear it from our room. Not tonight though. Actually, that's probably why I couldn't sleep. It's so quiet.'

'I've never known you to be such a light sleeper,' Dong Yiyao said.

'I've always been a light sleeper. Even when I was a baby, according to Yeye at least. I'd wake up from a cat meowing outside.'

Dong Yiyao studied his daughter as she began folding the paper into a crane. In the past year, her previously childlike face had opened up like a flower, and she was beginning to resemble her mother more. When was the last time he looked at her so closely? On most occasions, he would only see her as she was rushing to school, or when he was on his way out to work. Some days, after a long night, he'd sleep in and wouldn't even get to see his daughters in the morning. On his days off, when he did manage to make time for them, he'd take them swimming, fishing, or even on a trip to the police station to let them see his work. Now that he thought about it, he rarely ever made a point to sit down at a table and take a look at his daughters, and see how they were no longer who he thought they were.

Dong Ji had grown up. He felt surprised at that revelation. Though they looked nothing alike, she reminded him of Amu sitting in the passenger seat of the car and going on about finding a way to live for ever out in the desert. He wondered whether Amu was happy now that she'd abandoned her mother and lost herself. She couldn't have been happy. Could she even feel happiness? Could she feel anything? Watching Dong Ji's concentrated expression, he was reminded of his wife. The hand that was in his pocket clenched into a fist.

'Yeye told me that I should start thinking about what I want to do after I finish school,' Dong Ji said, extracting Dong Yiyao from his thoughts.

'He wants you to take over the pharmacy,' he said.

'Yeah. But I'm not going to listen to him.'

'So what do you want to do?'

'I don't know. I just don't want to do what he tells me to do.'

'Look at me,' Dong Yiyao said, his voice serious. His daughter complied.

'You need to know,' he said. 'You have to find a purpose. Something to work towards.'

'Yeah, I know. You're always saying that.'

'It doesn't have to be the pharmacy,' he said, 'but you've got to find something. Without a purpose, you will have no responsibilities, and that is not the way to live.'

'Like what though?' Dong Ji asked. 'What's your purpose?'

It should've been the simplest question in the world, but Dong Yiyao found that he did not know what to say. Dong Ji waited with anticipation, expecting him to provide an answer.

'Duty,' he said. 'Towards you, your sister and your Yeye.'

'That's rather vague,' Dong Ji said.

Her tone flitted ambiguously between satisfaction and disappointment. He wanted to know which one it was, but she did not tilt one way or the other. He waited some more, to no avail.

'Do you think the sun will be all right?' she asked instead.

Her question caught him by surprise. He had not seen Dong Ji fret over the sun before. In fact, his father had told him that, compared to the other children, his daughters had both taken the news of the sliver of the sun going away rather well.

'You'll be all right,' Dong Yiyao said.

'You don't have to say things just to make me feel better. If the sun's not OK, we're going to have a problem.'

'There's nothing to be afraid of,' he said.

'Like I said,' she said, yawning. 'I'd rather you just be honest with me.'

'Listen to me,' he said. 'No matter what happens, no matter how unsure you are, don't think about how you can change things. Don't think about making your own sun. When we try to understand something like the sun, we will always lose.'

Dong Ji shot him a sceptical look as her eyes narrowed, like a pedestrian who couldn't decide whether the person walking towards her was someone she knew.

'Have you been drinking?' Dong Ji said. 'You don't sound like yourself at all. Make my own sun? I can't just make a fire and call it a sun, Ba. Don't worry.'

'Just remember what I told you,' Dong Yiyao said. 'Just put your head down and live your life without thinking about things that are outside of our control.'

Dong Ji seemed like she did not know what to do with this warning. She shrugged and nodded at the same time.

'You're starting to sound like Yeye,' she said. 'Maybe you're getting old.'

'Whatever happens,' Dong Yiyao said, 'I will make sure that no one will harm you.'

'You've spent too much time with criminals, Ba. Not everybody is out to harm us.'

Dong Ji stood up and threw the paper crane into the garbage bin.

'I'm going to bed,' she said.

'Let me come in and say goodnight,' Dong Yiyao said.

'What?' she asked. 'Why? You never do that.'

But she seemed too tired for an answer. She poured herself a glass of water. Dong Yiyao followed her into his daughters' room. His youngest was asleep, her body turned to face the wall. The fan was on and she had kicked her blanket to the side. Dong Yiyao draped it over her and sat at the foot of her bed for a moment while Dong Ji placed her glass on the nightstand and climbed into her own bed.

'Don't wake her,' Dong Ji whispered.

Dong Yiyao gently squeezed his younger daughter's foot through the blanket, fighting his desire to rock her awake to hear her voice again. She would've fallen right back to sleep, but he did not make a sound and instead dawdled by her bedside, hoping that she would turn her body around. She did not move at all. In the end, he didn't even see her face.

Back in the living room, Dong Yiyao looked at the clock. It was just past two in the morning. Seeing that he still had time before the night was over, he took out Dong Ji's mother's old flute and began wiping it with a soft towel. He'd do this every once and again when he felt in need of her company. Somewhere in the back of his mind, he'd always held the belief that she would return. Even now, he'd placed his faith in her coming back one day. It'd been so many years. He should've given up long ago, but he couldn't stomach the thought of never knowing why she'd left and where she'd gone to, so he'd fabricated this hope that he'd grasped onto like a lifeline all these years. Without it, the loneliness would've been too much to bear. Now, he needed to believe in it more than ever.

While he was cleaning the keys, his old father came out of his room to use the toilet.

'Dong Ji was awake,' Dong Yiyao said. 'She's in bed now.'

'Have you eaten dinner? We have some leftovers.' His father headed towards the kitchen.

'Come sit with me, Ba,' he said. 'I'm not hungry.'

'Well, I am a little.'

The old man found a tin of red bean cakes that was by the television and sat across the table from Dong Yiyao.

'You shouldn't be eating such sweet things,' Dong Yiyao said.

'I know, I know. I'm cutting down. It's not easy for someone my age to change himself.'

His father placed a bag of tobacco on the table. The filters and papers were folded inside the bag. They each rolled a cigarette. The old man took puffs from his cigarette between bites from the cake. Under the dining-table light, his father's eyes were like the desert, Dong Yiyao thought, dry and yellow and endless. When Dong Yiyao was a child, his own eyes had always had a foggy quality to them. He'd often stare absent-mindedly at a single point in the distance until somebody interrupted him. According to his mother, it'd given him the unfortunate appearance of being less smart than he really was. He'd spent years of his childhood training himself to blink more, to move his eyeballs every time he caught himself in a daze. Just as he'd thought he had overcome this weakness, he'd see his reflection in the mirror again, brushing his teeth, with that blank look in his eyes. Years went by, he grew up, and just as he'd resigned to the thought that there was no changing himself, his mother told him that he had grown out of his habit. *You'll have no problem getting that job at the police station*, she'd told him, *you look smart now.*

As inconsequential as it may have sounded, it instilled in him the unshakeable belief that he was capable of altering his own fate.

'I quit my job,' Dong Yiyao told his father.

A shocked and worried expression rose to the old man's face. He put out his cigarette and moved himself to the chair beside his son.

'What happened?' the old man asked.

'I suppose I felt like I didn't deserve it,' Dong Yiyao said. 'I don't know. I'm just not sure I'm really cut out for this job.'

'I don't know if anyone is really cut out for the job of a policeman,' his father said. 'It's a tough life. Especially these days, with the sun and all.'

'Sure it's tough,' Dong Yiyao said. 'But I'd like to think that I've managed pretty well all these years. You know, there have been plenty of cases where we were too late to save the victims, but I'd always known where we'd gone wrong.'

As Dong Yiyao spoke, that old fear gushed up inside his heart. It was more debilitating than it'd ever been before. The feeling was the same one that had made him question whether he'd ever be able to reset his gaze, overcome his weakness and find a way to win with the hand he'd been dealt. It was what he'd felt every day looking at his clouded eyes in the mirror, only now it was pushing him down with a force much stronger, as though it would not stop until his bones were crushed and shattered.

'I'd always known that if things were done differently,' Dong Yiyao continued, 'those people could've been saved. But this time, no matter how I play the scenarios over in my head, I can't think of what I could've done to save them.'

'You can't save everyone, son,' his father said. 'You can't even save me from myself.'

He pointed to the tin of red bean cakes and did his best to smile.

'You've saved enough people,' he added.

'Do you really think so?' Dong Yiyao looked at his old man, whose expression was more troubled than assuring. 'Do you think I can save the girls? Save you?'

'You've lost me. What is it that we need to be saved from?'

His father inched his chair towards Dong Yiyao and placed his hand on his son's back. His warm and firm touch made Dong Yiyao feel like a child again, his back narrow and small, and his father's hand like a hard shell protecting him.

'Take this,' Dong Yiyao said as he fished out the folded photograph from his pocket. He stuffed it inside his father's palm and bent his father's fingers over it.

'Keep it,' Dong Yiyao said, his voice slipping out

weakly. 'Look at it when you're alone. Hide it. Don't let anyone else see it.'

'Did you get yourself into trouble?' his father asked. 'What is this? I knew you shouldn't have worked for the police. I knew it when you were a kid. It's just not who you are.'

'It is everything I am, Ba.'

'So what is this?'

'Something that nobody should ever have to see again.'

'Why don't you just throw it away?'

'Because it's evidence, I guess.' Dong Yiyao gave a feeble smile.

'I told your mother,' his father said, looking down at the folded photograph with furrowed brows, 'to be a policeman, you have to be able to accept your limits. There are times when bad people will win and there is nothing you can do. You have to be able to live with that, son. You have to be able to go home at the end of the day and forget about the job. Otherwise, you'll kill yourself.'

Dong Yiyao had heard words like these from his father throughout his life. The way his father cast a judgemental light onto his decisions had always ticked him off. He'd taken it as his father trying to coax him into quitting his job and inheriting the pharmacy instead. When Dong Yiyao was younger, he'd engage in heated arguments with his father, calling him manipulative and incapable of real love. As the years piled up, Dong Yiyao would try his best to walk away whenever his father mentioned anything about his career choices. Never once had he considered that his father might be right. Even now, his father's words made him angry. He had the qualities of a good policeman, he was certain of that. If it weren't for his persistence, so many cases would've been left unresolved and so much evil would've gone unpunished.

'I know that there are things I can't do,' Dong Yiyao said. 'I'm not a fool, I know that I can't correct everything rotten in this world, but that doesn't mean I should accept it.' Dong Yiyao stood up. 'A good policeman should never accept what's so clearly wrong,' he said. 'He should keep trying until the end.'

His father brought the cigarette between his wrinkled lips. The tip glowed red for a moment and then dimmed again.

'You are but a man, Yiyao.'

'Then what is the point?'

'I guess for most people,' said his father, 'the point is to be happy.'

'Happy?' Dong Yiyao felt ridiculous. 'Are you happy?'

'Yes, I would say that I'm happy.'

'Even when Ma was dying? When she was dying a little more every day, in front of your eyes, and you couldn't save her. Even then, you were happy?'

The old man put the rest of his red bean cake back into the tin and brushed the crumbs off his shirt.

'The sooner you accept things that can't be changed,'

he said, 'the sooner you'll be free from yourself. I know you're upset, but now I'm glad you quit.'

Dong Yiyao had considered destroying the photograph, knowing very well it was what he should've done, but somewhere inside him lay a kernel of desire: the will to protect the truth above all else.

He gave the photograph to his father because he knew his father well. All his life, his father had avoided getting into any trouble. He was consistent. Even when he'd once witnessed a theft on the farm he'd been working at, he'd hidden himself from the thief and waited for someone else to report the incident. Dong Yiyao knew that his father would not go to the police. He'd hide the photograph to the best of his ability. His father had always believed that keeping a low profile was the only way to guarantee a peaceful life, and a peaceful life was all that his father had ever wanted.

'I want my daughters to grow up,' Dong Yiyao said. 'That's all. I want them to live their lives and grow old. Like you did. That's what will make me happy.'

'Of course they will,' his father said. 'You don't have to worry about that. They're strong girls.'

'I'll make sure of it.'

'Now you'll have more time to take care of them. It's a miraculous thing, watching them grow. You'll see.'

His father pushed the tin of red bean cakes towards Dong Yiyao and told him that he should try one. He shouldn't go to bed on an empty stomach, his father cautioned, before retreating into his room to sort through the accounts for the previous day.

Dong Yiyao packed the flute back into its box and raised it carefully to the top of his closet. He fished out the paper crane from the garbage bin and unfolded it. On the paper was a list of professions: doctor, singer, chef, police officer, university teacher, news anchor, restaurant owner. He'd forgotten until now that a few weeks ago, at the breakfast table, when Dong Ji had told him that she did not know what she wanted to do after finishing school, he'd asked her to come up with such a list. Together, he'd promised her, they would narrow it down.

To the best of his ability, he folded the paper back into a crane and lowered it into the garbage bin.

Back in the field, Dong Yiyao opened the car door and saw Amu lying on her side, in the same position he'd left her, as though time hadn't passed for her. He had not needed to tie her up and he felt terrible for having treated her like a criminal. He swallowed the dry and hard lump of air that had been lodged inside his throat and moved swiftly to untie her hands and feet.

Without delay, he drove towards the lake. Amu's light was like a fire burning behind Dong Yiyao's back. A few times, he felt like he was going to pass out from the heat.

The sky had taken on a deep indigo hue. The night was slipping past him. He parked his car next to the lake and turned off the engine.

'Is this what you wanted?' Dong Yiyao asked.

As expected, there was no response from Amu. He considered once again that maybe it was all a nightmare, but he'd seen Ma Gang and Amu with his own eyes, felt their heat on his skin. He even had photographic evidence. It was the hard truth, as real as the fact that the sun was going to rise the next morning. Yet as much as Dong Yiyao believed in the necessity of truths, it was clear that he had only ever wanted innocence – ignorance, even – for those he loved.

Dong Yiyao could not apologise to Amu. There was nothing an apology could do for her. In its stead, he offered her the promise that her life had been meaningful. She was gone, he knew, she had forfeited her life the moment the light gobbled up her head. What was left was only proof of her foolishness, of the tolls she had to pay for yearning for the impossible. How could this be an eternal life? When Amu's lips parted and the light came out, her mouth had been like an open wound.

He held Amu's hand and led her to the pier and she followed obediently like a lost child. The lake reflected the entirety of Five Poems Lake back to itself like a giant mirror. But the image it showed was quieter and colder than reality, like the water had erased all the people who lived there, reflecting only the things that were frozen still.

Dong Yiyao heard a soft gasp behind him. When he turned around, he saw the Su girl standing about ten metres or so away. He just now remembered that the Su siblings slept on the boat on some nights. The girl covered her mouth with her hands and stared, wide-eyed, at Amu. Her body was positioned as though she was going to run away at any moment, but she didn't move an inch. Dong Yiyao couldn't tell if it was because she was paralysed with fear. Then he remembered that she had a bad leg.

'Is that where the sun disappeared to?' the little girl asked. 'Did she come from the sun?'

'She's sick,' Dong Yiyao said, trying to hide the worry in his voice. 'She's just sick. That's all.'

'I should tell the police,' the girl said, taking a step backwards.

'I'm a police officer,' Dong Yiyao said. 'You better go home.'

'I'm staying on the boat tonight,' she said. 'With my brother.'

'Where's your brother?' Dong Yiyao looked towards the boat. He could see nobody else around.

'He's asleep,' the girl said. 'Are you going to go to the hospital?'

'What?'

'You said she's sick.'

'That's right,' he said. 'She is sick. Now you go back to your boat, or I will arrest you for interfering in police business.'

He knew that the Su family had been in distress as of

late. The girl's father had run away with all their family's savings. Though his wife chose not to come to the police with the case, the Eastern District was a small one, and gossip often spread like pollen. His wife knew that all the money was under her husband's name; the police wouldn't have been able to do anything even if she were to report the theft.

Upon hearing Dong Yiyao's threat, the girl began to sob.

'Don't tell Ma,' she muttered. 'She would cry again. When she cries, she doesn't stop.'

Dong Yiyao made the Su girl promise never to speak of the events of that night, to which she obeyed and hobbled towards the boat as quickly as her legs could take her.

Once he'd made sure that the girl was back on the boat, he led Amu farther down the rim of the lake until the boat was no longer visible. How could he have forgotten about the Su siblings? Before all of this, he never would've neglected such an important detail. Thankfully, it was only a girl, at an age where other memories would soon overwrite this one. In time, Dong Yiyao and Amu would become a vague memory for her, if a memory at all. Even if she were to tell others, nobody would believe her.

Dawn's breath closed in like that of a killer. Amu seemed like she no longer felt any pain, like a breath of life had been blown into her spine. No, Dong Yiyao reminded himself, she was no longer Amu. She had been ill, yes, but that pain was proof of her having been alive. Now it was gone and so was she. He had done all he could for her. He had nothing left to offer but his life. It was his final act of duty to stay with her, to take responsibility for the things he knew and the people he'd failed to save.

He was sorry, he wished she knew. He really was.

BEACONS EVERYWHERE IN FIVE POEMS LAKE

11

The Su girl couldn't remember everything. All she knew was that Ba was at the pier with a Beacon. She was almost certain that the Beacon was a girl, though the more she thought about it, the less she believed in her own memories. Growing up, her brother often avoided going home, so most nights, he'd take her with him to stay on the boat. He couldn't stand being in the presence of their mother. He'd been too young to know how to confront their mother's deteriorating mind and body, so he'd avoided her entirely.

The Su girl did not tell her brother about the policeman with the girl who had a glowing head. She did not tell anybody. We asked her whether Big Su had been asleep at the time, to which she pressed on her temples with her thumbs and said that she could not recall seeing him leave his bunk bed. Then again, she couldn't remember whether she even checked.

The ambiguity of her memories frustrated us. The more I listened, the more I developed the urge to go swimming. I'd never been particularly good at it and, out of all the children around me, I'd been the least inclined to spend my days splashing about in the lake. But I was beginning to feel frightened of the truth that lay within the Su girl's memories, and I found myself thinking about swimming, being naked in water, at once vulnerable and impermeable. In a way, we are strongest when we rid ourselves of our clothes and our armour, when we expose our skin to the cold.

'What happened afterwards?' Dong Ji asked.

'I don't remember,' the Su girl said. 'I went to sleep, I think.'

'That's it?' Dong Ji said.

'What else do you want to hear? Your father told me to keep my mouth shut, so I did. What else do you want from me?'

'Why didn't you tell us?' I said. 'When the Beacons started showing up, why didn't you say something? Why didn't you go to the police?'

'I wasn't sure what to do,' the Su girl said. 'I liked talking about basketball with you. I just didn't want to bring it up.'

'You liked talking about basketball?'

Though I didn't mean to, I realised that I had raised my voice. I knew that I couldn't blame the Su girl for what had happened to our father. She didn't have to tell us about Ba, yet she chose to anyway. I should've been thankful, not angry. I lowered myself onto the sofa.

'Liu Mu was a girl, wasn't she?' Dong Ji asked Gao Shuang.

Gao Shuang nodded. He hadn't spoken a word since he'd brought Big Su inside.

'Who's Liu Mu?' I asked.

'Liu Mu is the university student that Ba was looking for,' Dong Ji explained. 'Ba ... provided false information regarding witnesses, I think, to have the case transferred to him. Liu Mu disappeared from the hospital on the day before Ba died. So I just wonder whether the Beacon with Ba could've been this Liu Mu.'

Gao Shuang continued to nod while Dong Ji spoke like it was only natural that she'd hidden this information from me.

'How do you know about this?' I asked. I couldn't believe that Dong Ji possessed knowledge that I didn't have. 'How long have you known?'

'I don't know,' she said. 'Gao Shuang told me a few years ago.'

'Why am I the only one who doesn't get told anything?' I said, standing up. 'How is that fair?'

'I didn't think it'd change anything,' Dong Ji said. 'I thought it'd only bring up painful memories for—'

'That's for me to decide!' I shouted.

'It's all in the past,' Dong Ji said. 'Right? I really didn't think it'd matter.'

'You're the one who cares the most about the past!'

Silence loomed. Light from the Beacons shone

through our curtains. The linen took on the translucent blue of butterfly wings.

I couldn't understand why I had to be protected, why the youngest must always be the child, and why all the people around me believed in the illusion that it was a privilege to live without knowledge. I couldn't see how ignorance could ever be the safest state of existence. I was mad at Ba for leaving behind so many questions, frustrated at Yeye for taking all his feelings with him into his grave, and fed up with Dong Ji for trying to pretend like she could assume the roles of every missing member of this family. All of their decisions were senseless and outrageous to me. Did they think they could breathe for me after I'm dead?

In spite of all my feelings, I couldn't tell Dong Ji that she would never be able to really protect me. It'd break her heart, render her sacrifices pointless and foolish. Perhaps we can only be whole when we can stop feeling responsible for another person's happiness. Until then, we will continue to rip ourselves into shreds and give those shreds away in the name of love.

I sat at the dining table. Dong Ji put her hand on my shoulder.

'I know what you're going to say,' I said.

'I wasn't going to say anything.'

'When I was little,' the Su girl said after some silence, 'I read a story about ghosts coming back to haunt their families. It was clearly supposed to be scary. You know, the mother that floats in the dark, the father that lurks in the garden, that kind of stuff.'

She ran her fingers through her brother's hair.

'It doesn't make sense, you know?' she continued. 'Why would that be frightening? I wish Ma would haunt me like a ghost, but she haunts me in ways that are loving and clings onto my memory as if she were real. It must be the same for you and your father. I think I can understand.'

She took a deep inhale as though she was pulling all the air around her into her small body. She let it out slowly through her nose.

'I'm not sure anybody can really understand,' Dong Ji said.

All of a sudden, we heard a thundering knock on the back door and an imposing voice bellowing from the other side.

'Anybody home?' a man called out. I didn't recognise his voice.

At first, none of us moved. After making sure the curtains were drawn, Dong Ji told Gao Shuang to take the Su siblings into the pharmacy, which he executed swiftly. She reminded me to say nothing unless absolutely necessary before she opened the door.

Two police officers stood outside with stern faces and towering frames. They looked at Dong Ji first, and then stretched their necks to check the rest of the room.

'Have you seen Gao Shuang?' one of the officers asked.

Dong Ji shook her head resolutely, as though she had been anticipating the question and had readied her response beforehand.

'We've been told that he comes here quite a bit,' the officer said, unconvinced by Dong Ji's answer.

Though I didn't know all the officers in the Eastern District, I recognised most of their faces. I hadn't seen these two before.

'We know he's here,' the officer continued. 'We have a witness who saw him here. If you're hiding him, there will be severe consequences.'

Most likely, they were bluffing. It was a common way of getting somebody to talk. Ba had often voiced his distaste for this method of intimidating, deceiving and manipulating others into professing their wrongdoings.

'Did he commit a crime?' I asked.

'That's none of your concern,' he said. 'We need to come inside and take a look.'

Dong Ji moved out of the way immediately, as though she had nothing to hide. It made me nervous. I had to stop myself from making any unintended movements or glances towards the pharmacy.

'Thank you for cooperating,' the other officer said. He was younger, his tone was kinder, and he seemed slightly embarrassed of his partner.

The two officers checked the rooms separately while Dong Ji and I followed without a word. They were swift and rehearsed in their movements and I was relieved to see that they were respectful of the place. I watched as the older officer stopped in front of the altar. He stared right at the placard, which had the photograph hidden behind, and I clasped my hands together to stop them from shaking. To my surprise, he did not touch the altar but instead, he turned his torso, stiffened his legs and raised his hand to his temple for a salute. Then, he let his body collapse to its fatigued state and let out a scarcely audible sigh before moving on, failing to notice the photograph. Once they had made sure that all the bedrooms were empty, they headed towards the pharmacy.

The older officer brushed the quilted curtain aside and opened the door. Neither of us could think of something to say in time to stop him. I watched, my heart racing, as he poked his head into the shop. The curtain fell on his back, blocking the view from us. A few seconds later, he recoiled his head and closed the door.

'It's a nice pharmacy,' he said.

Surprised by his remark, and relieved to know that Gao Shuang had somehow avoided being discovered, all I could do was nod. I quickly thanked him and said that, unfortunately, it had been looted during the protests.

'Dong Yiyao chose to become a police officer over inheriting this nice little pharmacy?' he asked. 'Seems like this would've given him a more peaceful life.'

He let out another regretful sigh and glanced towards the altar.

'Report to the station immediately if Gao Shuang

contacts you,' he said. 'It's very important.' He gestured to the younger officer and together they stepped out through the back door.

'I can't say he made the right choice,' he said before heading off. 'But for what it's worth, your father's contributions were very important to this town.'

Gao Shuang and the Su siblings had been crouching just behind the counter. Had the officer done a more thorough inspection of the pharmacy – had he even stepped foot inside and walked around for a few seconds – he would've discovered them. We didn't know why he stopped where he did. Perhaps it was out of respect for our father, or maybe he simply couldn't be bothered. We were all so relieved that the reason did not matter.

We ate more congee for dinner. Or was it lunch? I couldn't keep track any more. Unexpectedly, the gas was still on, so we cooked some of the vegetables that we had and pickled some of the rest. We knew they weren't going to last us long, but none of the markets were open any more. For now, we did our best to live in the present, focus our attention on the food we were eating, the heat on our skin, the weariness rising to our heads. I'd never considered just how impossible it would be to sleep under such bright light. None of us even managed to take a nap.

Dong Ji and Gao Shuang ate on the sofa while the Su girl and I sat by the dining table. Big Su refused to eat, so after some deliberation, we gave him a chair and found a space for him in the corner of the room. Every few minutes, Dong Ji would interrogate Big Su about the Beacon he'd seen. Not once did he talk. Eventually the Su girl decided that it was too much and asked Dong Ji to stop.

'Don't force him,' the Su girl said. She sounded unsure, like she was asking a question rather than making a demand.

Dong Ji did not stop; I wasn't even sure that she'd heard the Su girl. She held Big Su's shoulders and lowered her eyes to match his. I stood a few steps away from him, though I hadn't been avoiding him intentionally. He seemed different from the time I'd met him at the station, when he ridiculed Miss Pan's life and spoke about her death like it had been small. I remembered sensing so much distance between myself and this man, as though we hadn't grown up in the same town, eaten the same crops, breathed the same air, bathed in the same sunlight. I'd been angry then – there was no doubt about that – yet my anger had been muffled by a more disorienting feeling of disharmony and isolation. Back then, I had questioned whether it was really possible for one human to fail so entirely to find something recognisable in another. That feeling, to my surprise, was no longer there.

Locking eyes, Dong Ji continued to fire questions at Big Su. I watched the Su girl's hesitance give way to desperation. Her expression hardened, like the skin on her face was turning into stone.

'He's not going to talk, can't you see?' the Su girl said. 'He's confused. He's lost. He thinks he's a Beacon! Can't you see that he's not well?'

Dong Ji heard her this time.

'But he's not a Beacon, is he?' Dong Ji responded.

'He's sick,' the Su girl said. 'Can't you tell he's sick? He doesn't know anything. I was the only one there. Even if he did, he wouldn't be able to tell you. Just look at him. We were only children back then—'

'Being a child is no excuse.' Dong Ji stood up. 'What about all those years you could've told us, but chose not to? Were you still a child? When will we all stop using innocence as an excuse?'

Gao Shuang got up from the sofa, but before he could do anything else, Dong Ji shouted, 'We were children too!'

She stabbed her index finger at her own chest.

'We were innocent!' Dong Ji continued. 'We lost our father! And you were the last person to see him alive. We didn't leave him behind like you two left your mother. We didn't run away. You slept on that boat every night, you think I don't know? But we stayed right here. In this house. We waited for him to come home and he never did. All my life, in this freezing, blackened city, I've had to live with the consequences of that.'

My lips wouldn't move. I didn't know what to say to calm her, or if I even wanted to. I understood how she felt. We were sisters, after all. Every one of her words, like pieces of a puzzle, fitted themselves into the cavities within my heart. I couldn't help but wonder whether our bond was built upon the innate love between sisters or whether it was some sort of common sadness – from being motherless and then fatherless – that brought us closer to one another. Perhaps nothing was ever so clear-cut, let alone the inner workings of a family.

The best I could do was grab Dong Ji's elbow, though I didn't pull her back or tell her to stop. The Su girl pushed her body between Dong Ji and Big Su and stretched out her arms to block her brother from the rest of us. She proceeded to let out a cry – a guttural, frantic kind of roar that made her seem more animal than human. It came out of nowhere, but I had the feeling that it had been there this whole time. We all fell silent as she summoned this enormous roar from her chest. I saw in her a person who was at the edge of her body. It made me feel like a sun could burst out of her mouth. I instinctively took a step back.

'It's Dong Yiyao's fault!' she cried out. 'If he'd let me go to the police, then none of this would've happened! I've tried to be kind, to be sympathetic, to let the past go. But if he didn't stop me, maybe none of these Beacons would be here today, and my brother wouldn't think that he's one of them. Maybe he would be able to sleep and eat properly. It's killing him. He's going to die. And it's all Dong Yiyao's fault.'

'Our father is dead,' Dong Ji said. 'Whatever you say, he paid with his life. We don't have anything to do with it.'

'My brother is my only family now.' The Su girl lowered her body and buried her face in her brother's chest. 'Ba left us. Ma died. I can't let anything happen to him. I don't know why he's like this.'

Big Su's eyes looked straight ahead, as though he didn't even notice his sister burrowing herself into him. The Su girl noticed his indifference and bit down on her lower lip. I watched her gaze travel towards us as her youthful features contorted with hostility. I thought I saw all the muscles in her small body tense up, like a famished animal protecting its prey. It was as though all her innocence had burned to dust and was blown away by a swift gush of wind. Innocence is such a fragile thing. What is the purpose of being born with a quality so pure, only for it to be replaced one day by something tainted and ugly?

'It's all Dong Yiyao's fault,' she said. She talked through her teeth.

With an unhinged look in her eyes, she pushed herself away from her brother and headed straight for the altar. Her lame leg was like a sandbag weighing her down as she limped across the living room. She picked up the wooden placard and smashed it on the corner of the wall.

Dong Ji and I scrambled to rescue the placard from her hands, but she was stronger than she let on. She grabbed the roots of my hair and yanked at them. A blade-sharp pain pierced into my scalp and threaded down my spine. Gao Shuang rushed over and grabbed the Su girl's arms, restraining her. The Su girl tried to shake him off, to no avail. Just as I thought it was over, I saw her knee Dong Ji in the torso with her good leg, sending Dong Ji falling into the altar. The apple rolled off and knocked over the glass of baijiu on its way. The clear liquor spilt onto the photograph. Dong Ji fell to the ground in pain and Gao Shuang released the Su girl and rushed over to check on Dong Ji. I wanted to grab the photograph, but before I could move, the Su girl smashed the placard again on the altar.

The wooden placard broke into pieces. Ba's name was split right down the middle and the photograph lay wet on the floor. For a moment, I must've been the only one who noticed the photograph. I stopped. We all did.

'That is enough!' Gao Shuang yelled at the Su girl.

Despite all that had happened – the shattered placard, the pain still fresh in my scalp, the altar in disarray – I still couldn't believe the hostility I was seeing on the Su girl's face. I couldn't imagine that she was the same girl I'd so carelessly conversed with just a few days ago. I never imagined myself to be so naive. Or perhaps I'd never thought the world could be so deceptive.

I started to pick up the wooden pieces. With each piece, the image of Yeye's hunched body sitting alone at his desk grew more vivid in my mind. I saw him carving the golden strokes, his hands trembling slightly. What had he been thinking?

'We should be ashamed,' I said. 'I'm so ashamed.'

Dong Ji looked away and her eyes rested on the wet photograph on the floor. Gao Shuang saw it then too. He picked it up and stared at it for a while, as though he recognised something.

'You're right,' Dong Ji said to the Su girl. 'If you'd gone to the police that night, maybe none of this would've happened. Maybe we'd still have a father. Maybe they could've saved him. I guess, in that sense, we're even.'

Dong Ji's voice had lost its vigour and taken on the quality of dust on the ground, like a fire in her had died and only ashes remained.

I should've said that blaming one another was not going to make any of this easier. It was unfair, not to mention unwise, to jump to conclusions. It would only rip us apart even further. I knew it was not the Su girl who was at fault. I should've told them that nothing would get resolved this way. But I kept my mouth shut because, just like the rest of them, I was angry. Even though I couldn't bring myself to accuse the Su girl of failing to save Ba, I admit that it brought me some respite to hear Dong Ji say it.

The temperature of the room had risen rather drastically now that more Beacons had congregated in this area. There were only hushed, vague noises outside, like the sounds of early mornings when everybody is curled under their blankets: sounds that were silent. The Beacons wandered the streets and sat in corners. Some were even lying down, like they were asleep. Of course, there was no way to tell whether that was true.

'He deserved it,' the Su girl said quietly, like a blanket had been draped over her voice. 'He was not a good police officer.'

For a moment, it looked as though Dong Ji was going to spit shards of glass out of her mouth, but Gao Shuang spoke before her.

'This is him,' he said, still looking down at the wet photograph. 'I'm sure of it. This is Ma Gang! We used to make fun of that scrawny body of his.'

All of us turned towards Gao Shuang.

'You know him?' I asked.

'He didn't seem like the type,' Gao Shuang said. 'I can't believe he had it in him. All that heat and light. Captain Dong must never have given up on that case.'

'He didn't seem like the type?' Dong Ji said. 'How are we ever going to find out anything if that is the best we can come up with?'

'I don't know,' Gao Shuang said.

'I'm tired of hearing that,' Dong Ji said. 'I'm sick of not knowing anything.'

Gao Shuang set the photograph down on the altar, poured a glass of water for himself and gulped down the entirety of it. It trickled down the corners of his mouth and onto his navy T-shirt, dyeing the fabric an even darker shade. No matter what he felt, I knew that he never would have argued with Dong Ji. It was his way of loving her; ensuring that he did not add to her burdens, regardless of the sadness it caused himself. He wanted to be the one to help her take a breath when she was suffocating. He would let her stand for ever on his shoulders, even if it meant that he wouldn't see anything.

My scalp was still in pain. Dong Ji and I both had scratches on our faces. One of Dong Ji's cheeks was beginning to bruise.

Big Su gazed out the window with a blank look. The bowl of congee that'd been on his lap fell and spilt onto the floor.

'Fuck,' Gao Shuang said. He grabbed a towel that was hanging on the kitchen sink faucet and squatted down to clean up the spill. 'Sorry.'

Dong Ji paid no heed to the shattered bowl.

'None of this changes anything,' she said. 'I need him to talk.'

Before these past two weeks, she'd rarely ever articulated what she wanted for herself. Her reasons had always been for the sake of me, or us. It caught me off guard for a moment, but almost immediately, she corrected herself.

'We need him to talk,' she said. 'All of us. We've been living with so many questions.'

I assessed Big Su. He didn't seem like he cared at all. For a moment, Dong Ji was deep in thought, her expression unreadable. She held her breath while she deliberated and the whole room fell silent. She then exhaled loudly, letting out all the air that she'd held stale inside her lungs, and looked at me squarely in my eyes.

'I realise this is a strange thing to say,' she said determinedly.

She walked up to me, grabbed my hands and rubbed their backs with her thumbs.

'Let's burn the boat,' she said.

Big Su was the first to react. He looked at Dong Ji, whose eyes had already been prepared to meet his gaze. He didn't say anything, but the fact that her words elicited a reaction in him encouraged her even further.

'I won't do it if you tell us what you saw,' she said to him.

'I'm not doing it,' I said. 'And I can't let you do it.'

'We should talk this through,' Gao Shuang suggested.

His passivity was frustrating. Maybe we were all losing ourselves, growing new skins to adapt to the environment around us, becoming unrecognisable to those who had known us. Had I changed too, then? What did I look like now?

'I'll burn it myself then,' Dong Ji said.

After the fight, the Su girl seemed to have lost all her vitality. Unlike her brother, she did not seem to care about the boat at all. Her eyes were as dead as Big Su's were moments ago. She slumped into her brother's body like dough resting against a basket.

Dong Ji proceeded to search the living-room cabinet. She quickly found a sunhat that hadn't been used in years. She put it over Big Su's head and started dragging him towards the door.

I shot Gao Shuang a look, imploring him to do something to stop her, but he followed her, as though, in just a few seconds, he'd made up his mind as well.

I ran to the entrance of the pharmacy before Dong Ji could get there.

'Ba wouldn't want this,' I said.

'Ba is dead!' Dong Ji said.

'If you really believe that, then it shouldn't matter any more.'

'How can you say something like that?' Dong Ji's voice was shaking a little. 'Nothing matters more.'

'What about the other possibility?' I asked. 'The possibility that what we've been too afraid to say has been the correct answer all along. Maybe the Beacons don't matter. Maybe Ba just took his own life, Beacon or not.'

Dong Ji stared at me, like she'd been betrayed. There was no doubt that we'd both held the same speculations throughout the years, yet neither of us had said those words until now. To think that the father we'd looked up to and depended on would choose to leave us behind was something that neither of us could ever accept.

She pushed me aside and pulled Big Su out onto the street and towards the pier. The sunhat fell off Big Su's head but Dong Ji did not stop. The Su girl had tried to follow us, but she couldn't keep up. A breeze was rocking the white boat from side to side.

Dong Ji stopped in front of the boat. I realised that she had no way of burning it. She didn't have fuel. I looked around. There must've been at least a hundred Beacons around, none of them paying us any mind. Among them, I saw two police officers running somewhere. In their rush, they didn't even look at us.

Just as my fear was beginning to subside, Gao Shuang jumped onto the boat, descended to the bottom deck, and emerged again with a container of fuel in his hand.

'Do you know what you are doing?' I shouted at him. 'This is a crime, Gao Shuang, and you're an officer!'

I was furious. Had they both lost their minds? Not only was he refusing to stop her, he had decided to support this impulsive, foolish idea. The worst part was that I couldn't understand why. Whose was the right way of loving somebody? His? Or mine?

Dong Ji took the fuel from him. She opened the lid, sniffed it and started pouring the liquid onto the boat.

'Stop,' Big Su said. 'Stop, stop!'

His voice first came out as a soft, broken whine. It then grew louder as he found more power in his vocal cords. As though his lifeless body had been woken up by his voice, he jumped onto the boat and snatched the fuel from Dong Ji. I followed him.

Dong Ji took out the lighter from her pocket and held it in front of Big Su's face.

'I'll burn it!' she said. 'I will!'

The Su girl had finally made her way onto the boat

but her legs gave out. She fell without much of a fight and did not even attempt to stand up.

'If you really think the truth is that important,' I said to Dong Ji, 'you can burn us all.'

For a moment, the breeze went away and the rocking stopped. All of them, basked in the light of the Beacons, seemed as though they were glowing. Even Big Su's dark birthmark was illuminated with a pale sheen. The texture of their skin and the angles on their faces were all blurry and pallid in the white light. I must've appeared the same. We must've all looked like ghosts to each other.

I stood in place, showing Dong Ji that I had no intention of leaving. She lowered onto one of the seats, clenching the lighter to her abdomen.

'Why is it that we don't know anything?' she said. 'How is it fair that our fates are dictated by things we know nothing about?'

'I know that you are my sister,' I said. 'The most important person. That's all I know and that's all that matters to me.'

She looked up at me with tired eyes, as though something had just fallen over inside her.

'That's not enough for me,' she said.

Even though I shouldn't have, I looked away from her. I couldn't bring myself to meet her eyes. I knew that it wasn't her intention, but it was impossible not to feel betrayed, angry and bitter. In the distance, I thought I saw someone I recognised standing by the pharmacy, but it was too bright to confirm.

In the corner of my eye, I thought I saw Big Su turn into a Beacon. I spun towards him and realised that I'd been imagining things. He was standing on the side of the boat, eyes fixed on something near the burned police station. I followed his gaze. The Beacons – hundreds of them – were gathering near the remains of the building. They continued walking until they were squeezing their bodies together and climbing on top of one another, stacking themselves into a shapeless mound. Their bodies vanished, one by one, into the collective light. None of us dared to move. I don't know how long we stayed on the boat, watching the Beacons congregate to form this giant, bright mass. It appeared that Beacons from other parts of Five Poems Lake were also being drawn here, as though they had been summoned by a voice only audible to them. We watched until the light grew taller than all the homes around it.

Big Su bent over and helped his sister up with both hands, as though her bones were soft and new, like those of a baby. As he did that, I thought I saw a mother in him.

'Dong Yiyao killed her,' Big Su muttered under his breath.

12

Big Su had not been asleep that night. He'd been swimming in the lake and he saw it all. He'd often done this, he recalled. The water calmed him those times he'd wake up from his nightmares. Even as a child, he did not sleep much. It must've been some sort of mental weakness he'd inherited from his mother, he said.

Big Su told us that the night Ba died, he'd seen Ba with what he'd thought was a miracle – a Beacon. According to him, when he saw them, they were standing by the lake, and Ba was holding both of her hands in his. His head was lowered, like he'd been apologising to her. Then, slowly, he turned around and took a step into the lake, dragging the Beacon in with him. They never came back up. By the time Big Su managed to swim closer to take a look, all he was able to see was a light distorted by the rippling water, merging with the reflection of the rising sun in the distance. Ever since that night, he couldn't take his mind off the shining brightness that he'd seen. He grew up and became a police officer, all so that he could find out what it was that he'd seen that night.

While I listened, all I could think about was what this information was going to do to Dong Ji. The whole time, I was too afraid to look at her, worried I might find out that the person I wanted to hold on to had been broken herself. In my current state, I would not have been able to piece her back together. If neither of us could be strong for one another, then what would happen to us?

'He was a police officer,' Dong Ji said, after Big Su was finished with his story. 'Not a murderer.'

Her voice sounded surprisingly calm, and when I heard her speak, all the dread and doubt that'd accumulated inside me vanished in a flash. I spun my head eagerly towards her, only to find out how wrong I'd been. Despite the steadiness in her voice, she did not look at all like herself. She had an expression that I'd never seen before. It wasn't so much an expression as a colour, like a darkness had been cast on her face and her skin had deepened into an obscure tint of grey and brown and green. It was the colour of time-worn, tarnished skin, like she'd grown into a different person right in front of my eyes.

'He was good,' she said, nodding her head as though to confirm her own words. 'He was a good man. Everybody knew that.'

Her eyes shot towards me like a dart, as a gesture for me to agree with her. Gao Shuang reached for her arm, but she raised her palm in the air.

'Don't,' she said. 'Don't, please. I don't want pity. I don't need it.'

Gao Shuang recoiled and addressed Big Su instead.

'Why didn't you tell the police?' he asked.

'Would anyone have believed me? Would you have believed me?'

'No,' Dong Ji interjected. 'Of course not. Because it's ludicrous to think that Ba would've done something like that. Maybe it was another officer you saw.'

I wanted to say something so that she could become herself again, but my mouth hung open, empty. I was so fixated on Dong Ji that I couldn't even begin to think about what our father had done. Unlike her, I dwelled not on the fact that he'd killed a Beacon, but that he'd taken his own life willingly. It seemed so foreign, so impossible based on what I remembered of the indestructible man that was our father. But he had been gone for twelve years, and I only knew him as a shadow of a father, a silhouette in the mist of my memories. Dong Ji was the one alive and breathing in front of me; she was the only family I had and the only one that was still real.

'So stop lying to me,' Dong Ji said to Big Su. 'Tell me the truth. Tell me what you really saw.'

'You wanted the truth and now you can't handle it?' Big Su said unconcernedly.

His eyes were still fixed on the tower of Beacons on the police station as they continued to fuse into a single light.

'What reason could he have had for doing that?' Dong Ji asked, looking at Gao Shuang and then at me. 'He would never have abandoned his family. His job was to save people, not kill them. He just wouldn't have hurt anybody. Why would he? He was good. He was the most righteous person I knew.'

Neither of us were able to answer her. The Su siblings ignored her as well. It was clear that they didn't care at all about us. Maybe this was what they'd wanted; maybe this was who they were. The way Big Su had talked about Miss Pan had made him seem so spiteful, and perhaps this was spite as well, revenge for Dong Ji having threatened to burn his boat, although it did not appear to have brought him any pleasure. He just gazed, rapt with desire, at the tower of Beacons, while his sister ran her palm along his spine as though she was stroking a kitten.

'OK,' said Dong Ji, like she'd suddenly thought of something. 'I can believe that there is no reason for Big Su to lie to us, but it was dark, right? And it was twelve years ago.'

She turned to address Big Su.

'There is just so much room for you to remember

incorrectly,' she said. 'If you say that you saw Ba and the Beacon, I'll believe you. But the details? Maybe you're wrong about the details. Have you ever thought that it could've been the other way around? The Beacon could've killed Ba.'

'They don't harm people,' Big Su said. 'They're just trying to help us.'

'You're out of your mind,' Dong Ji said coldly. 'You're obsessed with them, like they're some sort of god. What have they done to help us? Look around you. You said it yourself; you thought the Beacon was a miracle. What did you say? Something about it being the most beautiful thing you've seen? Is that why you're putting the blame on our father?'

The searing heat had made it impossible to breathe normally. The air burned my lungs with every inhale, like my head was over a boiling pot of water. My body was exhausted and brittle like a burnt matchstick and my legs felt as though they could send me collapsing at any moment. I didn't want to listen to them argue any more.

'Let's go back inside,' I said. 'We'll get a heatstroke out here.'

'Heatstroke?' Dong Ji asked with a sudden bite to her tone. 'Did you not hear him slandering Ba? Do you not care? He's saying that Ba killed someone!'

Without pause I shouted, 'He was my father too! I don't even have what you do! At least you saw him one last time! And now he's gone, Yeye's gone, and I hate this feeling too. But it is not all about you and what you want!'

The words exploded from me without any warning. When I suggested going back home, I had meant to approach carefully, in a non-threatening manner, and I hadn't felt a single trace of anger inside me, so I was just as stunned as Dong Ji was when those words spilt out of my mouth. If I hadn't even sensed this feeling, how could I have doused it before it erupted from me so uncontrollably?

Gao Shuang reached for Dong Ji's hand again. This time, she did not refuse.

'Big Su's right,' I said. 'You don't even want the truth. What you want, what you've always wanted, is just something to keep you going, all because you can't bear the thought of living an ordinary, insignificant life like mine. You think you have to go and chase after something so big that you don't even have the courage to look at it when it's in front of you. You can't just live to be alive. You always have to have something bigger, a purpose. And because of your decision, now I have to deal with a truth I never wanted to hear. Of course I care, Dong Ji, I care just as much as you do.'

I did not wait for her to respond. I turned around, climbed out of the boat and stormed towards the house. With every step I took, I resisted looking back. I did not want to see the expression on her face, nor did I want to show her mine.

Dong Ji and I had been through so much together, hand in hand for all these years. We couldn't possibly have expected that we'd experience all of it the same way, come out of it as the same person. Why, then, did we feel such disappointment when we couldn't understand each other? I couldn't tell whether I was weak, childish, or just human, to feel so lonely, to want to deny the fact that loneliness was only natural. On some days, it felt like I was all of those things. On other days, it was as though I was none of them.

    I believed Big Su's story, though admitting to this felt like a betrayal. To be honest, I wished that he'd never told us. I felt stupid, like we'd spent all this effort searching our bodies for an old scar that had already healed, just to cut into the same patch of skin again, all because we wanted to confirm that it did once exist, that there used to be a time when we were bleeding.

    We never should've dug up the urn. At the very least, the camellia plant might still have been alive. I didn't know why we couldn't just let death be. Before, when I watered the camellia plant, trimmed its branches, and watched the flowers blossom, when I dusted the altar, replaced the fruits, I'd been able to detect, however distantly, the elusive beauty in death and – sometimes – I'd even felt a flicker of joy. I'd found comfort in that feeling, more than I'd ever imagined.

Like a hole, the more we probe into death, the deeper and emptier it becomes. Now, Ba was really gone. All of him.

When Dong Ji came through the door, she was by herself.

'Gao Shuang is bringing the Su siblings to his home,' she said. 'He thinks you and I should have some time alone.'

Her voice was hushed. Her face had turned heavy. I could tell that she did not want to start another argument. Maybe it was because she lacked the energy, or maybe she felt that it was wrong of her to have accused me of not caring.

'He's afraid that the police will arrest Big Su,' she continued to explain. 'He really should worry about his own safety more. Big Su himself didn't even seem to care all that much.'

I'd drawn the curtains and switched on the ceiling fan to its highest speed. I'd found an old tank top and changed into it. Even indoors, the temperature was still hot enough to make me sweat, though it was much more bearable than outside in the light. It'd been so long since I felt sweat on my skin. The sensation made me feel like I was in somebody else's body.

'Look,' said Dong Ji. 'I didn't mean to direct my anger at you. I'm sorry—'

'Can you do me a favour?' I asked.

'What is it?' she said. 'Don't. I can't. You can't ask me to believe what he said.'

'Can you give me one of those baths?'

A glint of curiosity restored some life into her dull eyes.

'What?' she asked. 'A bath? In this heat?'

'I want one of those that you give people at the parlour.'

She sat down next to me on the sofa. I looked at my bare legs. They were so pale and the skin had not even the illusion of lustre. I sensed Dong Ji studying my face as she attempted to discern my thoughts. Then she seemed to give up and let out a long sigh that curved into a smile.

'Fine,' she said, standing up. 'Though it won't be the same. I don't have the stuff we use at the parlour, but I'll see what I can pull together from the pharmacy.'

She asked me to fill half of the tub with hot water while she prepared. She went into the kitchen and found a jar of coarse salt. I waited in the bathroom, by the wooden tub, while she picked out some dried lavender, aged tangerine peels, cinnamon bark, dried catnip and angelica roots, and combined most of it in a gauze bag.

'What about the lavender?' I asked.

'That's for something else,' she said, handing the gauze bag to me. 'Put this in the hot water and soak it for fifteen minutes.'

I took off my tank top and sat naked on a plastic chair. I pulled my knees to my chest and rested my left cheek against them. The steam fogged up the bathroom, obscuring everything. Feeling a little light-headed, I closed my eyes and bathed in the beads of sweat that were running down my face and body. If I were to faint, so be it. The floor was there to catch me. I melted into the misty heat.

When I opened my eyes to check the water, I saw that it'd taken on a barely visible tinge of orange. The smells of fruits and spices were pleasant. I wondered why when cooked into a pot of medicine, all these distinct flavours transformed into a uniformly woody and bitter taste. Perhaps it was because the bitter flavours always overpowered the rest.

Once I thought the fifteen minutes were up, I turned on the cold water and the steam cleared up almost immediately. It took a while for the tub to fill to the top. I called out to Dong Ji when the water was ready and she entered the bathroom with a bowl in her hands. She told me that she had mixed the salt with the dried lavender and some canola oil.

'Go on,' she said. 'Hop in.'

'Just like that?' I asked. 'I don't have to do anything?'

She reached one of her hands into the water to check the temperature.

'Good,' she said. 'You didn't make it too hot. I don't want you to pass out.'

'I didn't know what temperature it should be.'

'This is good. Get in slowly. Don't slip.'

I stepped on the chair and climbed over the tub. Embarrassed to let her see my dried and cracked skin, I hurried to lower myself into the water.

'I gave you so many baths when you were young,' Dong Ji said. 'You hated them all. Why did you want one now?'

'I did? I hated them?'

'Yeah,' she said. 'Sit up taller. Turn your back towards me.'

She picked up my hair and held it to the crown of my head with a hairclip that'd been clasped onto her sleeve.

'You just always wanted to do other things,' she said. 'I had to make sure you brushed your teeth every day too.'

She took some of the salt mixture from the bowl and smeared it over my shoulder. Her fingers glided across my skin in fluid movements. She must've done this hundreds of times before. The salt scratched against my skin and I wondered why anyone would find enjoyment or relaxation in this. It was more painful than anything.

Dong Ji must've noticed my discomfort, and I felt her reduce the pressure and slow down her movements.

'Why do people pay for this?' I asked.

She laughed. 'Your guess is as good as mine,' she said. 'But I'm glad they do. Otherwise, I'd be out of a job.'

'It hurts,' I said. 'My skin feels like it's peeling off.'

'That's the point. The salt removes dead skin. Makes you look younger, like a baby.'

'My skin will never look like a baby,' I said. 'Are people really so stupid? They think that by peeling off their skin, they can reverse time? Do they think some salt and lavender will enable them to be reborn somehow?'

She laughed again. 'Just shut your mouth and enjoy,' she said. 'Where do you get all these questions from?'

'It's not enjoyable,' I said. 'That's my point.'

'Well,' she said. 'Think about how expensive this would be normally at the wellness parlour. Maybe that'll help you find enjoyment. Plus, didn't you want this?'

The lukewarm water had cooled my body a little. Dong Ji lifted my right arm out of the water and rested it on the edge of the wooden tub. She ran her palms from the back of my hand all the way up my arm and started rubbing in little circular motions on my shoulder. The salt rolled beneath her hands. When I submerged my arm back under the water, some of the oil and lavender floated to the top.

'I wonder what's going on outside,' I said.

Dong Ji kept quiet as she started on my other arm and rubbed the crevices between my fingers. All that could be heard were the sounds of my legs in the water as I crossed one over the other.

'Let's leave,' Dong Ji finally said. 'Let's leave today.

Once Gao Shuang is back, let's all get out of here together.'

'So the two of you have already decided?'

I didn't know what she'd say if I refused to go with her. Would she go anyway? It had always been an unthinkable decision, something that could theoretically happen but in reality never would – an act akin to a mother leaving her child behind. But how absurd that assumption was. Weren't we both left behind by our own mothers?

I couldn't see how it was possible to live comfortably while loving somebody else. Where does one find the courage to face the possibility of such a loss? Sometimes, I didn't know whether there really was a difference between love and dependence.

'If you've decided,' I said, 'why are you asking me?'

'Do you really want to stay here, in this place, where all we have left are painful memories?'

Her hands stopped for a moment as she let her question hang in the air.

I had never liked Five Poems Lake. That much was true. But not once had I had the desire to leave. The memories were not all so bad, I wanted to tell her. In fact, to me, they were the only good things about this place.

I tried to caution myself against allowing sentimentality to cloud my judgement. Logically speaking, there was no reason for me to stay. The most important thing was to stay together now, I told myself, no matter where we go. Dong Ji had always been the one to make a decision when I dithered, and I'd always listened to her. It could very well be that I was just afraid to leave this familiar life behind, to venture too far from home. I wasn't like Dong Ji. I had to dig for the determination that she so easily found inside her.

'Trust me,' Dong Ji said.

She was right. I should trust her. She always knew better.

Making her hand into a cup, she scooped water onto my other arm.

'We'll make it,' she said. 'We won't die. We'll take Gao Shuang's car and drive it as far as we can.'

'Do you think the farms can grow crops again?' I said. 'Now that the Beacons—'

'The Beacons won't save us, little sister, like how we can't quench our thirst by swallowing our own spit.'

She told me to raise my legs out of the water. She began scrubbing the soles of my feet.

'What about Ba?' I asked. 'What about Yeye?'

'What about them?' she asked. 'Ba was a policeman and Yeye was a pharmacist. They're both dead. Nothing has changed.'

'But who will take care of them?'

'There's nothing to care for,' she said. 'A wooden placard and a headstone are just those things. Wood and stone. That's all. If you want, we can bring the placard with us.'

I withdrew my leg and told her it was enough.

'I can't tell a difference,' I said, looking at my calf now red from the rubbing. 'It just hurts.'

Dong Ji rinsed her hands in the bathwater.

'You have to get used to it,' she said.

I pulled my knees to my chest and wrapped my arms around them.

'Ba always said that he loved us so much that he would die for us,' I said. 'Do you think that's why he died? For us?'

'Love is not dying for someone,' Dong Ji said as she wiped her hands with a towel. 'It's trying your hardest to live for them.'

I pulled the clip out of my hair. The damp strands hung down my shoulders and floated on top of the water.

'Clean the salt off,' she said. 'And then let's get everything prepared before Gao Shuang comes back.'

We combined the pickled vegetables into two large jars. We boiled the rest of our eggs and decided that we'd add them to the pickling liquid. In the freezer, I had a bag of cured goose liver sausages that I'd purchased from the market a few weeks ago. Despite trying our best to bring all that we could, we had to leave some food behind.

'We'll eat the eggs first,' Dong Ji said while she peeled them. 'Save the sausages.'

'How much rice can we carry?' I asked. 'We have three and a half bags.'

'Bring three.'

'Three bags? That's heavy.'

'Bring it, just in case.'

I lugged the rice out of the pantry and onto the table where we'd piled our supplies. We'd packed two sets of outfits each, five large bottles of water, a small pot, a flashlight, some candles, a knife, some basic medicinal herbs and one large blanket.

'We don't have enough water to last a week,' I said. 'We don't have enough empty bottles. Should I go and check if the convenience store is open? That man would do anything for money. Although, in this heat, I don't know.'

Dong Ji finished peeling the egg that she'd been working on and wiped her wet hands on her T-shirt. She came to the sofa and looked outside the back window. From where we were, we couldn't see the Beacons, but the entire Eastern District was bathed in their light. I followed her out the front door to the pharmacy.

It was brighter than ever. It seemed like most of the Beacons had made it to the remains of the police station. The mound had grown into a gigantic size that seemed as tall as the buildings in the Western District. It felt as though we'd walked too close to a fire. The mound of light was moving a little, and parts of it were sticking out here and there as it continued to morph into an oddly shaped mass. Feeling my skin start to burn, I pulled Dong Ji back inside. Another Beacon, still dressed in her winter coat and knee-high boots, walked by our door and in the direction of the light.

I looked through the glass door that faced across the lake. One of the buildings in the Western District had an angled surface that reflected the light into the sky like it was a signal to something above. The building glowed gold and beautiful. The lake was teal, but where the light hit the strongest, the water had become shimmering white, as though the surface was powdered with silver dust. It was difficult to believe that, just a few days ago, it seemed as though the snow was never going to stop.

'We'll never know what's happening to this world,' Dong Ji said. 'How could we possibly know?'

Even though I heard her, I'd been so entranced by the view in front of me that I didn't realise she was speaking to me.

'Just like how we'll never know why he died,' she said.

This time, her words jerked me back to reality. I should've felt relieved to hear her say it, but strangely, I felt nothing, as though I didn't care about all of that any more. I meditated on whether it was really possible to live in harmony with this light. Looking at Five Poems Lake in its current state almost made me believe that the mound of light was really going to replace the sun. But even if I were to allow myself this hope, I knew that there was no guarantee the light would do anything to help us. We didn't know whether it would nourish our soil, feed our plants, strengthen our bones.

We didn't know anything, I agreed with her: we'd never known anything.

But what was different now was that I felt hopeful. It was a hope that most certainly did not exist out in the darkness of the desert. It was the kind of hope that only light could bring.

'It's Gao Shuang,' Dong Ji said, pointing at a black car that was approaching. 'Come on. Let's finish packing.'

Already, she'd spun around and dashed back into the living room, beckoning to me as she walked. I stalled and watched Gao Shuang park his car the way people did when they didn't expect to stay long – at an angle, without any careful manoeuvring. Just a single turn into the kerb and a step on the brake.

'I can't drive with this light in my eyes,' he said. 'I can hardly keep them open. Where's Dong Ji?'

I pointed towards the living area.

'She's packing,' I said.

'Does she want to go right now?'

Gao Shuang was about to take a step towards the living room when he stopped himself.

'You're coming, right?' he asked.

I looked around at the pharmacy, which was still disorderly and in need of at least another day of cleaning up.

'Right?' Gao Shuang asked again.

I think I nodded. My eyes were busy assessing the apothecary cabinet. A few of the drawers were open.

'Good,' he said as he placed his large hand on my shoulder, turning me around so that together we would walk towards the living area.

'She wasn't sure she could convince you,' he said. 'I'm glad she managed.'

'What about you? What about your parents?'

'They have my brother,' he responded instantly as though he'd anticipated my question, or rather, he'd already asked himself the same question and arrived at the answer.

'I'm needed more by you two,' he added.

We lifted the quilted curtain aside and entered the living room, where Dong Ji was standing with her back to us, shoving supplies into a backpack with all her weight. Gao Shuang walked up to help her.

'I brought some bottles of water,' Gao Shuang said.

'Did you fill up the tank?' she asked him.

'We have enough to get to the edge of Five Poems Lake.'

'You didn't fill it up all the way?' she asked.

'We won't need that much fuel.'

'You're right,' she said. She looked at Gao Shuang and then at me. 'You're right. I'm sorry. I'm just nervous.'

I tried to see whether I could detect any doubt in Gao Shuang. It was hard to believe that Big Su's story did not shake his heart in any way. I realised that whatever feelings he had were overshadowed by his guilt for the pain he'd caused Dong Ji by bringing the Su siblings here. A few strands of his hair, wet from sweat, stuck to his forehead. He interacted with Dong Ji with so much care, like she was a stray that would run away if he uttered a word too loudly, or a flower that would wither if he poked at it with too much force.

Just as I was thinking this, Dong Ji said to Gao Shuang, 'How are you doing? You must be shaken too.'

'I'll be all right,' he said calmly. 'I don't believe Big Su either.'

'You don't have to say that for my sake,' Dong Ji said. 'I'm not a girl you have to protect any more.'

She went up to the altar and grabbed the broken placard that we'd taped back together. She brought it with her to the dining table and tried to find a space for it in the backpack. There was none, that much was clear.

'Leave it,' I said. 'Let Ba be.'

She must've thought that I hadn't really meant it, because again she tried to fit the placard into the side pocket of the backpack, but it was too large. She blew gently into the carved characters and, having rid it of dust, returned the placard to the centre of the altar.

Dong Ji took the front seat of the car and I got in the back. Gao Shuang drove slowly and vigilantly towards the desert in the north. As we distanced ourselves from home, the light became gentler on our eyes, allowing Gao Shuang to gradually speed up the car. There were people on the roads, many travelling towards the Eastern District to take a look at the source of the light. To my surprise, a few buses drove past us. I didn't know that the buses were still running. I imagined that Red Bean must be at work too. There was so much to report on, after all. I wondered what her panel of experts had to say about all of this. I could imagine their perplexed faces, and the image brought me a sense of comfort and familiarity.

We did not see a single Beacon on our way. They must've all flocked to the Eastern District and merged themselves with the light. Even from far out near the farms, we were able to see the top half of the light glowing above the buildings.

As we drove, I couldn't stop thinking about the green apple on the altar. Had it been chipped or bruised during our struggle with the Su girl? If so, then the exposed flesh would begin to attract all sorts of insects. I imagined the peel turning brown, the fruit shrivelling up and worms nibbling away at it. Were there worms left in Five Poems Lake? I hadn't even thought about checking the apple or replacing it with a new one.

There was no point fretting about it, I reminded myself, we were miles away from home now. Even if the fruit was intact, it was bound to rot sooner or later, especially now that Five Poems Lake had become so hot.

While I worried over the apple, Dong Ji talked about all the what-ifs and how-tos of our imminent journey into the desert. I heard her call my name.

'Did you hear me?' she asked. 'If we see anyone else, ignore them. We don't want trouble.'

I slid into the middle seat. Through the front windscreen, I surveyed the vast skies ahead of us. I had never been this far out of town. Everything was flat and, from underneath the miles of cracked soil, dead trees stuck out like splinters in dried skin. I could see where the colour of the clouds deepened, which must've marked the farthest distance the light could reach. In another twenty minutes or so, we'd be driving into darkness. The wind coming in through the windows was growing colder. I moved back to the side seat and rolled mine up.

Dong Ji turned around to check on the light in the distance.

'Look at it,' Dong Ji said. 'It's still growing taller. At this rate, it's going to burn everything. The sun is supposed to be in the sky, not on the ground. This thing is going to destroy the town. I say it's a good thing that we got out.'

Then, to reaffirm her decision, she said, 'With the three of us together, it'll be like we never left home, like we just moved it around and changed it up a little bit.'

I looked at the upper half of Gao Shuang's face through the rearview mirror. He kept his gaze on the road and I noticed him moving his hands restlessly, sliding them up and down the wheel and changing his grip every few seconds.

Could homes be moved and reassembled? What about the water marks on the ceilings, the warped wood, the smells soaked into the furniture? How could we have brought that with us? What about the lives that were lived within them, the words spoken and bounced off the walls, the people who added and subtracted to the place? How could we possibly pick all of that up and drop it somewhere else, expecting it to be the same?

And what about us? How much of myself was I leaving behind? Back when I turned eighteen and Dong Ji moved out, weren't there traces of her that'd stayed? I'd seen those very pieces of her scattered around the house, in the organisation of the refrigerator, the scents of the soaps she'd used, her chair by the dining table, the doorknob that she'd replaced. We were not just our bodies.

'I'm sorry,' I said. 'Stop the car. I'm sorry.'

'What's wrong?' Gao Shuang asked. 'Do you need to use the toilet?'

'I'm not going. I can't go. I don't want to go.'

Dong Ji spun around. Her eyes grew empty. Gao Shuang turned his head quickly towards Dong Ji and then reverted his attention back towards the road again. He continued to drive, though I felt the car slow down a little.

'Stop the car,' I said again.

'Don't be difficult now,' Dong Ji said, spinning her body back to face the front. 'This is not the time.'

I pulled on Gao Shuang's T-shirt.

'Let's stop and talk this through,' Gao Shuang said as he turned the wheel and the car came to a halt by the side of the road.

I opened the door and pushed myself out. They both did too. Outside, the wind had grown stronger. It dried the sweat on my back, sending a shiver through my body. I felt my skin stiffen, like somebody was pulling on it.

'I've watched over our home all my life,' I said, raising the volume of my voice so that I could be heard through the wind. 'It's all we have left of our family.'

'*We* are all we have left of our family,' Dong Ji said. '*You*,' she pointed to me and then to herself, 'and *me*! If you go back there by yourself, you won't have anything left but bricks and wood. I don't understand you.'

I knew it wasn't that she didn't understand me, or that I didn't understand her. But it is a near impossible thing to do, to give up on changing someone you love.

'We can't leave you alone,' Gao Shuang said. 'Remember when we were having lunch together a few days ago and I said that once the sun is gone, we should all leave together?'

'And remember I said that it wouldn't matter where we were if the sun really disappeared?' I responded. 'I was wrong. It does matter. It matters where we live, and it matters where we die. If I were to leave, then I'd no longer be myself.'

'Can you do it for me?' Dong Ji said. 'Just this once? I've never asked you for anything.'

'I already do everything for you,' I said. 'Everything I've done is so that you can be happy. You've never asked, but I've always been here to give. Now, I want to stay for myself.'

'We've always talked about doing this together,' she said. 'Why did you grow up to be so selfish?'

The wind blew strands of loose hair across her face, like large brushstrokes of black ink smearing over her features.

'I never wanted to leave,' I said. 'Why does that make me selfish?'

'You never once told me that!' she said as she swiped her hair back, revealing her reddened eyes. 'How was I supposed to know?'

'I didn't want to add to your burdens,' I said. 'I didn't want to disappoint you. And I really did think that I could leave with you. I didn't want to let you go alone.'

'You never have to trouble yourself for me,' she said, looking away and wiping her eyes with her fingers. 'I can take care of myself.'

'Why can't you see that, by saying exactly what you just did, you're the one who is being selfish? How can you ask me not to care? How is that not selfish?'

The same twisting and aching feeling that had propelled me to shout at her on the boat had imposed itself onto me again. I thought it'd gone away, but there it was in my chest, hiding behind gentler intentions like a wolf behind a bush, silent and undetectable until it was too late to escape.

'Let's go,' Dong Ji said to Gao Shuang as she turned back towards the car.

'We can't leave her here alone,' he said.

'That's what she wants, isn't it? To be alone. Let her be then. She's old enough to decide for herself.'

I watched as she opened the car door and climbed inside, slamming it shut behind her. Gao Shuang grabbed my wrist and tried to lead me back into the car, but I pulled my arm from his grip, making it clear that I had no intention to follow him.

'Don't be so headstrong,' he said. 'She's just trying to protect you. That's what older siblings do.'

'What about your little brother that you're leaving behind? Haven't you realised by now? We can't protect one another from anything.'

'Then what the hell am I doing here,' he asked, 'if not to protect you two?'

He took a few steps towards the barren land and stood by the very edge of the road. I saw his shoulders rise and fall as he took five or six deep breaths before he finally turned around again.

'At least say goodbye properly,' he said. 'Otherwise, you'll regret it.'

He walked up to the passenger-seat side of the car and pulled on the door. It was locked. He knocked on the window and told Dong Ji to open it, but the door did not budge. He opened the back door and leaned his body inside.

I turned around and walked in the direction we came. I'd seen a bus stop not too far down the road – maybe a half-hour trek from where we'd stopped. I heard Gao Shuang call to me from behind. I turned around and told him it was all right, though I wasn't sure he heard me through the wind. I didn't know what to say to Dong Ji anyway. Maybe this was better, I thought, the only way we could let go of each other.

Everything that made me who I was existed in Five Poems Lake: its wet heat, its vanished sun, its endless snow, its blackened trees, its waters, its Beacons. I was a part of the town. I was its limb. I couldn't just detach myself and leave.

I would be lying if I said that I didn't immediately feel any regret for what I'd just done, that I didn't question whether my decision was really the right one. I was terrified of the idea of being alone, but what was worse was the thought that, had I left with them, I wouldn't have been able to find myself again, and another kind of loneliness would ultimately take me.

Then, as I realised that there exists a moment in our lives when all of us cease to change, I also understood that I, too, was not going to change any more. Fate had me seized tightly within its palms. I could never live independently of Dong Ji, yet at the same time, we were separate people. It was the very paradox that lay at the root of our love, our pain and everything else in between.

When I turned around again, the black seven-seater had driven off. I walked towards the bus stop, trying not to pay attention to the silence surrounding me. I wasn't ready just yet. I didn't want to think about the aftermath, about what tomorrow was going to look like.

I didn't want to admit to this, because I wanted to think that it was courage, or my sense of responsibility for our home, or some other virtuous reason that'd propelled me to stay, but I knew that somewhere in my subconscious, I was afraid of seeing Dong Ji die. I wanted to believe that she'd make it safely somewhere, even if it meant that I had to be left behind. I must've become conditioned to living with the unknown, rather than the hard truth.

The bus arrived late, though it did not matter. For a while I was the only passenger. I took a seat in the front row and watched the bus driver wipe his sweat as he drove. I wondered whether he had a wife and children and what kind of a man he must've been to be working on a day like this one.

The bus carried me towards the lake. As we approached, I saw that the tower of Beacons was beginning to resemble a giant, glowing person. For most of the ride, all I could see were the head and shoulders of this figure. Once we arrived at the last stop, which was directly across the lake from New Hope Hospital, I confirmed that the tower had really moulded into the shape of a human. One of its legs was planted firmly in front of the police station and the other one was standing on the Eastern Pier. I walked along the rim of the lake in the direction of home. The heat became so strong that I thought I could smell burning. I overheard a young woman calling it the 'New Light'. I didn't know whether she was the one who'd come up with the name. People continued to gather around the lake, as though they were waiting for something.

Just as I crossed into the Eastern District, I saw three boys swimming in the lake. They kept their squinted eyes on the New Light and pointed at it every so often as they treaded and glided in the water, which wrapped around their bare skin like translucent silk. I noticed that one of the arms, so to speak, of the New Light was slightly shorter than the other. Compared to the rest of the body, the light around that arm was stirring and flickering more intensely. It must've been in the process of finishing its transformation.

As I got closer to the pharmacy, I discovered that the leg of the New Light was blocking the entrance, though it was so bright that I couldn't even see the outline of our house. Unlike in the Northern District, nobody was outside in this area. It was too hot to be so close to the New Light for more than a few minutes. I circled around to the back alley, shielding my eyes with my forearm as I guided myself with the wall towards the house. I heard somebody call out to me from their window.

'It's right in front of you!' they said. 'The door is right in front!'

I should've thanked them – maybe I did – but I could hardly keep myself steady. I managed to find the keyhole quickly enough and pushed my way into the refuge of my home.

Two days passed. The New Light had managed to even out its two arms and was standing in the same place, emitting steady beams. It stood straight and its arms hung down either side of its body. I couldn't tell which way it was facing, so I imagined that it was looking towards the lake. The New Light had taken on the same serene quality of the Beacons that it was built with, or, if you held a more romantic view of all of this, the Beacons that birthed it. For those distant villages far out in the west, it must've appeared that the world was frozen in eternal dawn.

I began feeling an intimacy with, or even a desire for, this immense yet quiet light. It was in no way a bodily feeling, but rather the kind of comfort that one gets from the divine, the sort that makes the heart feel like it is supported by something mighty. It made me think that the light might not have come from inside the bodies of the Beacons, but rather from somewhere high above the sky or deep below the lake.

I was eating lunch, I suppose, when I heard a knock on the back door. It was twelve o'clock, but I'd lost track of whether it was noon or midnight. It was impossible to sleep on a schedule, so I'd given up entirely and resorted to letting my body dictate whenever it was ready to go to bed. Between cleaning the pharmacy, cooking, eating and sleeping, I spent most of my time watching television. Red Bean continued to show up every day for a few hours. It seemed like she was the one that named it the New Light. After hearing her repeat it a few times, I learned to like the name. Naturally, nobody knew why the New Light had come to be, so everyone resolved to give up on thinking about its origins and focus instead on its implications. Every day, I waited for Red Bean to talk about the Beacons and whether they were still alive in there, but she rarely even used that word any more.

When I heard the knock, I'd just turned off the television, as Red Bean had ended her programme and the screen had faded to black. I opened the door and saw Dong Ji leaning against the wall, the backpack dangling from her hand, the bags of rice by her feet. At the entrance of the alley, I saw Gao Shuang's car drive away out of sight.

I let her inside and closed the door. She hurled the backpack onto the dining table. Then, she pulled out a chair, sat down, crossed her arms over the table and buried her face in the gap between her chest and her forearms. I heard her crying.

I poured her a glass of water and placed it on the table next to her. The light coming in through the linen curtains was soft against her hair, illuminating each strand with the colour of cinnamon. I found a comb and, standing behind her, I ran its teeth through her hair. As I worked through the knots, she lifted her head and leaned her body back into the chair. I made sure I wasn't pulling too hard as I combed, until all of it was untangled and spread like a hand fan down her back. Her hair was much better maintained than my own. Each strand, though thin, was strong and sprung back to life as soon as I'd finished combing.

I never asked her why she came back. There have been times when, while we reminisced about the day we tried to leave, I noticed her falling silent as she held her breath, eyes staring vaguely at something in front of her, hands fidgeting. During those moments, I'd always think that maybe she was going to say something, but each time all she did was let the air between us hold still for a moment before ultimately deciding that she couldn't utter those words.

It was most likely that her words would not have been able to accurately explain her reasons anyway. Those words, I suspected, had much to do with her pride, her denial, her coming to terms with the fact that neither of us were as free as we wanted to be.

For a few more days, the New Light stood tall and watched over our town. Then, as though suddenly beckoned by something, it began to walk slowly into the lake. As it submerged its body deeper inside the water, the lake began to glow and the New Light became less dazzling. We were finally able to look at it.

It really was beautiful, more so than the sun itself. Perhaps it was the idea that we'd made it ourselves, with our own bodies. Not mine, of course, but still I felt like a part of it was me, or that I'd come out of it in some way. I came to accept that I could very easily have become a Beacon. It was good fortune, I realised, that helped me escape that fate. There was nothing that made me better than those who'd turned into Beacons, no quality that made me more deserving of life. It was just plain old luck.

The entire town came out of their homes and watched from the rim of the lake, like ants around a bowl of honey. The New Light continued to descend until only its head was left afloat above the water. Shortly after, the head, as though it had given all of us one last good look, followed its body and sank. The lake continued to shine golden for a while until the remnants of the New Light were extinguished, and the rest of us were left clinging on to some distant dream in the dark, cold chill of night.

We all went back home. I didn't know what was different, what the New Light did to us, whether it gave us hope or despair. I wasn't even sure whether it really existed. I went to bed in the dark for the first time in days. Dong Ji climbed into her own bed, which was pushed against the wall on the other side of the small room. Back when I was a child, every night, I would ask her to stretch her arm out and then I would do the same and meet hers in the middle. For years, I held on to her hand while I slept.

I flipped my body around to face her side of the room and noticed, in the dim streetlight that streamed in through the window, that she was facing me too. Her eyes were closed. One of her hands was resting under her cheek.

'Ba came in to see you,' she said in a mellow voice, her eyes still closed. 'The night before he died, he sat by the foot of your bed and pulled the blanket over your body.'

She opened her eyes halfway.

'I told him not to wake you,' she said. 'I know it's not my fault, but I feel responsible. I just want to tell you that he did come to say goodbye. I don't want you to blame him.'

I felt a pull inside me. It was not a strong one, just something tugging ever so slightly at me, like somebody was ringing a bell.

'What did you two talk about that night?' I asked.

'To be honest,' she said, 'I don't really remember. Probably nothing important.'

I turned to lie on my back and placed both my hands over my stomach. The warmth of my palms gradually penetrated through my shirt and into my skin. I heard Dong Ji shift under her blanket.

'Well,' she said, 'I guess I remember him telling me that I must never try and make my own sun.'

'And what did you say?'

'I think I told him that he was drunk.'

I heard her softly chuckle at her own words. I smiled too.

'He liked drinking,' I said. 'A little too much maybe.'

'They all did,' she said. 'The whole police force.'

'Everyone in this town does. They drink like they think the lake is made out of liquor.'

'Not us,' she said.

'Yeah.'

'Maybe we should,' she said, which made me laugh.

'Yeah, maybe.'

For some time, only our slow breathing could be heard. I wondered whether the moon was out there in the sky, and if so, had it been hiding in plain sight this whole time? Maybe I should've got out of bed to take a look at the sky, but I did not want to move. I was comfortable. I was where I was supposed to be.

'What should we do about this?' Dong Ji asked in a whisper, as though she too did not wish to disrupt the hush of the night.

'About what?' I asked.

'Everything.'

'I don't know,' I said. 'Not much we can do, I guess.'

Again we fell silent.

After a bit, Dong Ji asked, 'Do you want me to hold your hand?'

'I'm not a child any more.'

'I know,' she said. 'Still, let's hold hands. It'll help me sleep.'

And so we slept, each in our own beds, like two pages of an open book, our clasped hands hanging in the space between. The sensation of her palm touching mine was different from my memory of it. Her hand was so soft, like she was a child and I was her mother. My mind wandered for a while, through all kinds of sceneries, until it came to rest on an endless forest, with trees so old and tall that I couldn't even see how high they extended. There was no one in the image, only birds; little colourful ones that flew from tree to tree. Sunlight trickled through the foliage and onto the forest floor, and I continued to watch the birds without any kind of recognisable feeling, as though I had become just another tree myself. Time passed and I watched the forest grow and grow until I awoke to a warmth on my cheek and the morning light shining in my eyes.

About the Author

An Yu was born and raised in Beijing. She left at the age of eighteen to study in New York City. A graduate of the NYU MFA in Creative Writing, she writes her fiction in English and lives in Hong Kong. She is the author of the critically acclaimed novels *Braised Pork* and *Ghost Music*.